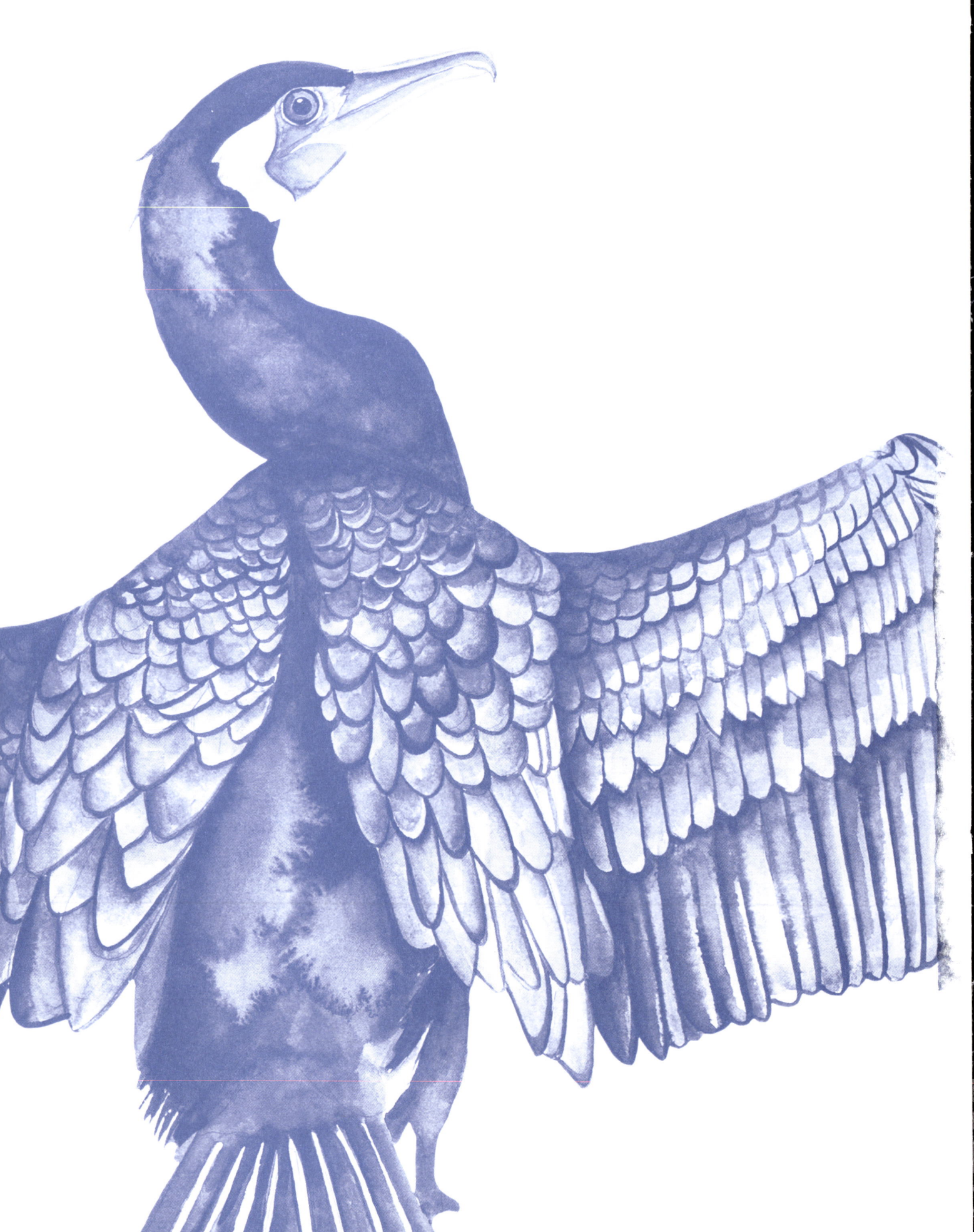

The Illuminated Book of

BIRDS

✳

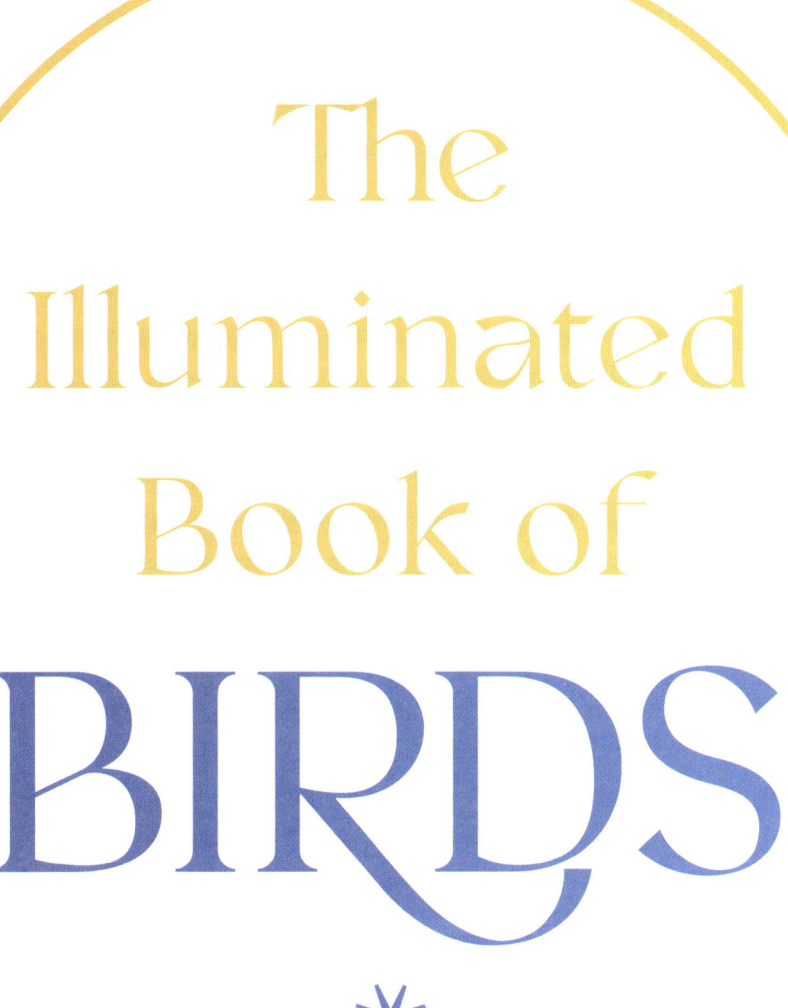

The Illuminated Book of BIRDS

Written & Illustrated by

ROBIN CROFUT-BRITTINGHAM

Timber Press ✳ Portland, Oregon

For Felix and Jake,
who I hope always live in a
world filled with birdsong.

Timber Press
Workman Publishing
Hachette Book Group, Inc.
1290 Avenue of the Americas
New York, New York 10104
timberpress.com

Timber Press is an imprint of Workman Publishing, a
division of Hachette Book Group, Inc. The Timber Press
name and logo are registered trademarks of Hachette
Book Group, Inc.

Printed in Dongguan, China (TLF) on responsibly
sourced paper

Text and cover design by Hillary Caudle

The publisher is not responsible for websites (or their
content) that are not owned by the publisher.

ISBN 978-1-64326-504-9

A catalog record for this book is available from the Library
of Congress.

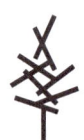

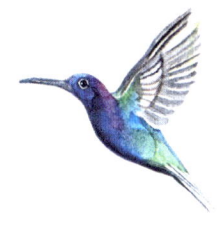

In order to see birds, it is necessary
to become part of the silence.

-Robert Wilson-Lynd

You can't really look for birds; you
can't make a bird come out and
identify itself to you. The most you can
do is walk quietly and wait until you
hear something, then stand motionless
under a tree using your animal senses
to figure out where and what it is.

--Jenny Odell, HOW TO DO NOTHING

Table of
Contents

✳

BIRDS of the WORLD

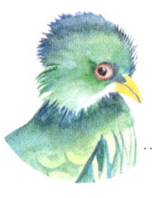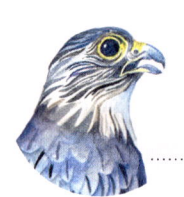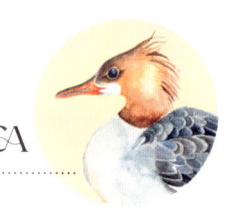

BIRD RELATIONSHIPS

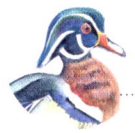 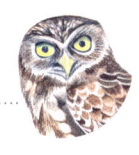
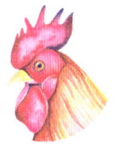 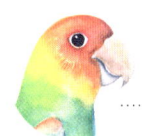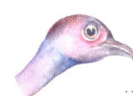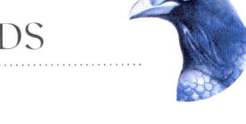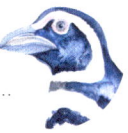

Preface

Birds have always captivated and beguiled humanity. Be it the glint of an iridescent feather catching the sunlight as a grackle flutters up to its nest or the gentle coo of a Mourning Dove as we awaken, there exists an inexplicable allure surrounding these winged beings.

Birds have transcended mere reality to inhabit the realms of mythology, literature, and religion, ingraining themselves as profound symbols, guides, and messengers. The ancient Greeks described a voracious eagle descending from the heavens daily to feast on Prometheus's regenerating liver, along with the steadfast falcon who accompanied the god Apollo on his celestial exploits. Meanwhile, in Native American creation mythology, the world takes shape as a waterfowl plunges into the depths of the ocean, emerging with a sphere of earth that is sculpted into dry land.

Birds have traversed the tapestry of our collective imagination, kindling the fires of myth and faith. In this book, several other such narratives will be told. Additionally, the reader will see illustrations of birds I find most beautiful and fascinating. Some of these birds have graced my presence in the wild, their melodies etched into my memory, while others remain elusive, seen only through the pages of other books.

I love birds deeply and I hope to share and inspire my passion in others. Perhaps my interest in birds was inevitable when my parents bestowed upon me the name Robin, after the red-breasted herald that signals the arrival of spring in the northeastern United States, my birthplace. This enduring love for birds has accompanied me from my formative years, marked by the ethereal calls of loons during summers spent in the wilds of Northern Ontario, to my adult adventures that have taken me as far afield as South Africa and the southernmost reaches of Argentina, always looking at beautiful birds along the way.

My aspiration for this book is twofold: to captivate the seasoned birdwatcher and to ignite the passion of the budding naturalist. During a winter visit to my grandmother's home, my infant son watched, rapt in fascination, as a group of sparrows congregated at a window-side bird feeder beside his high chair. A dear friend and fellow bird enthusiast who was dining with us remarked, "Well, this is how it starts." My hope is that this book serves as an inaugural stepping stone—a conduit that bridges the seasoned admirer of birds with the nascent enthusiasm of a youngster who is just beginning to cultivate an interest.

I am, first and foremost, an artist, and it is important to emphasize that this is not a scientific work. Although this volume features birds from every corner of our planet, it is not intended to serve as a comprehensive guide. Such an endeavor would span hundreds, if not thousands, of pages, and numerous such books exist already. My aim is to illuminate a curated selection of birds, drawing from diverse regions across the globe. I have always loved to look for birds when I travel and that is what inspired me to make these geographic groupings of birds. The groupings do not necessarily represent specific biomes or endemic samples but are examples of birds you might find in a particular region—be they seasonal visitors, endemic species, or introduced or invasive birds. I am in debt to many more prolific and studied writers and scientific illustrators for the accuracy of their research and representation, both of which have informed and inspired the work in this book.

Introduction

Today, more than ten thousand living species of birds inhabit our planet, and more are being discovered every year. From the frigid reaches of frozen tundra to the dripping heat of twisting tropical rainforests, birds live in every climate on Earth. They can be hiding among the dense foliage of a tree in the forest or soaring high above a billowing grassland in search of prey.

Humans have observed and documented bird life since the dawn of their coexistence on Earth, dating back to petroglyphs from Northern Africa that depict birds being hunted by wild dogs. The study of bird behaviors dates to ancient Rome and has long been considered auspicious—a word derived from two Latin roots: *avis*, for "bird," and *spicere*, "to see." Modern ornithology has advanced far beyond the original ancient practice of studying the behaviors of birds to foretell the future, but this tradition of symbolic meaning is still fascinating and relevant in many cultures to this day. Birds appear in myths and legends around the world, whether as a symbol of life hatching from the egg, the phoenix rising from the ashes as the symbolic totem of a new beginning, or the winged messenger carrying news from distant lands or other planes.

Throughout history, humans have relied on birds for meat, for their feathers, and even for their excrement, which is used as a potent fertilizer in agriculture. There are also numerous examples, both past and present, of the domestication of various bird species to aid in human pursuits. For example, in early Chinese civilization, cormorants were captured and trained to dive with collars around their necks that enabled them to fish but prevented them from swallowing their catch.

Birds have also helped humans survive. The phrase "canary in a coal mine," which refers to an event or omen that forewarns of a coming crisis, derives from a real practice. In the early twentieth century, British coal miners were vulnerable to hard-to-detect levels of noxious gases in the mines where they worked. After it was discovered that canaries were affected by gases more quickly than the miners, it became a widespread practice to take canaries into the mines and watch them for signs of distress to indicate the air was no longer safe. The birds would often fall unconscious or dead when exposed to harmful levels of gas. Birds were eventually replaced with modern technological advances in gas detection, but they undoubtedly saved the lives of many miners.

Humans have also kept birds as pets, dating back to the time of Alexander the Great in the fourth century BC. Greek soldiers captured wild birds in India and returned home with what have come to be named Alexandrine Parakeets. Fascinated by the parrots' ability to mimic human speech, aristocrats in Europe and the Mediterranean region kept them as entertaining pets. As the fascination with parrots and parakeets grew, they became increasingly popular. It was soon considered a symbol of status and sophistication to keep attractive birds in cages inside the home.

Mammals carry their young in their wombs, but birds, like their dinosaur relatives, lay eggs. The size, color, number, and even the shape of the eggs can vary hugely depending on the species: the largest egg is laid by the ostrich and the smallest by the hummingbird. After a bird lays an egg, it must be incubated and kept warm and out of danger from animal predators. Inside the egg, a chick uses its sharp beak to pierce the shell and emerge into the world. Like birds themselves, eggs are crucial symbols in folklore and mythology, embodying themes of rebirth and life. Nests, the structures that birds build to house and protect their eggs and chicks, can be constructed from a variety of materials, including rocks and mud, animal fur, grass and sticks, and artificial materials discarded by humans. Nests are built in every habitat, from the cavities of trees in the forest to hollows carved in cacti in the desert; some waterbirds even build floating nests.

Although thousands of species are alive and thriving today, many birds are threatened because of habitat loss, industrialization, and climate change. In her seminal 1962 book *Silent Spring*, American biologist and conservationist Rachel Carson sounded the alarm on the environmental impacts of widespread pesticide use by conjuring the chilling parable of a future with no birdsong. The legacy of the work in the conservation movement was immeasurable, and the text drew widespread attention to the fragility of the natural world. To this day, her work is a reminder of how crucial birds are to our ecosystem and how vital it is to respect and preserve them and the habitats that enable them to thrive.

In this book, we'll touch on birds from around the world and explore some of the behaviors and eccentricities of specific birds and what makes them significant to various cultures and peoples. We will also visit and remember some of the birds that have been tragically lost as the result of human negligence and overconsumption.

BIRDS
of the
WORLD

From the highest craggy reaches of the Pyrenees mountains to the glistening lakes of the Canadian north, let us spread our wings and fly around the world in search of some of the most beautiful birds on Earth. Along our journey, we will explore miniature, delicate flycatchers and large, mighty birds of prey. We will meet birds who can survive in the sweltering heat of the desert and those that can withstand the hostile, whipping winds of the Arctic coast.

Pull back the leaves of the low-hanging branches and see if you can spot the gleam of feathers. Listen closely and try to hear the sweet melodies of birds so elusive you only hear their gentle coos or their resounding caws.

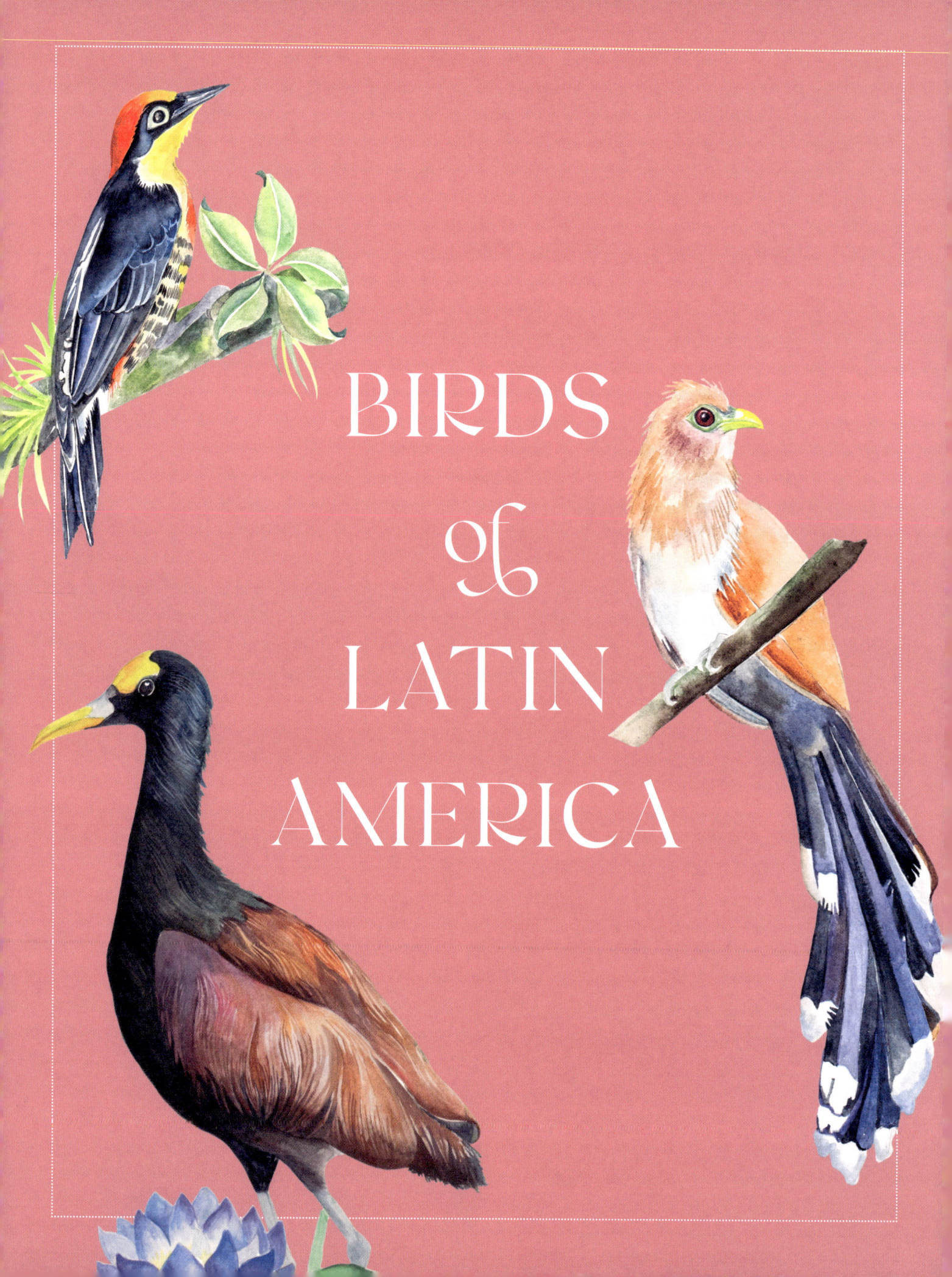

Birds

of

Latin

America

Latin America is rich with landscapes filled with riotous colors, bursting with flamboyant flowers, and teeming with poisonous plants and animals. Within these verdant vistas exist many diverse habitats for bird life, with thousands of species inhabiting this part of the world, making it one of the foremost destinations for birdwatchers and nature enthusiasts.

Look closely, and you'll spot the large rainbow-colored bill of the Keel-billed Toucan, believed by the ancient Aztecs to be sent by the gods. Keep your eyes peeled for the spectacular crown of the Atlantic Royal Flycatcher, and look for delicate, jewel-toned tanagers perched among the trees and looping vines. Soar along the coastline and spot the bizarre bulging red breast of the Magnificent Frigatebird, or see a Brown Pelican use his huge bill to scoop up fish from the sea.

BIRDS of COSTA RICA

With lush rainforests, creeping vines, and vibrant, otherworldly flora, Costa Rica is home to many extraordinary birds. Search among the rainforest canopy, between the draped leaves of towering trees and coiled vines, and you may spy one of the six species of toucan that live here. Though the Keel-billed Toucan (2) and the Collared Aracari (7) are relatively common, they can be difficult to spot, as they often dart among the thick leaves of the forest's canopy in search of succulent fruits to feast on.

The large bill of the Keel-billed Toucan may seem unwieldy, but because it is made of the protein keratin—the same material in human fingernails—it is quite lightweight. Ancient Aztecs believed that the rainbow coloring of their bills meant they were sent as messengers from the gods, and they worshipped the birds in hopes that the gods would send them rain. Although toucans are normally reserved and stay away from humans, they have been known to attack when they feel threatened, especially during mating season. In the early days of ornithological research, it was widespread practice for scientists to capture birds in their natural habitat and remove them for further study and research. Victorian naturalist Henry Wallace Bates was once attacked by a flock of toucans in the Amazon as he tried to capture one of the birds.

The Resplendent Quetzal (4) has also long held great cultural significance. The national bird of Guatemala, the quetzal is associated in Guatemalan culture with the deity of Quetzalcoatl, one of the most powerful and important gods of the Aztec pantheon. Quetzalcoatl is often depicted as a feathered serpent, with glimmering green tail feathers that represent the leaves of plants sprouting forth from the earth. The ancient Mayans and Aztecs considered the colorful feathers of the quetzal to be the highest symbol of status and fashion, and their feathers were often featured in the headdresses and diadems of emperors to show wealth and link them with the god Quetzalcoatl. As quetzal populations declined, the value of their feathers became so great that they were used in trade as currency. The official Guatemalan currency is named for the bird to this day.

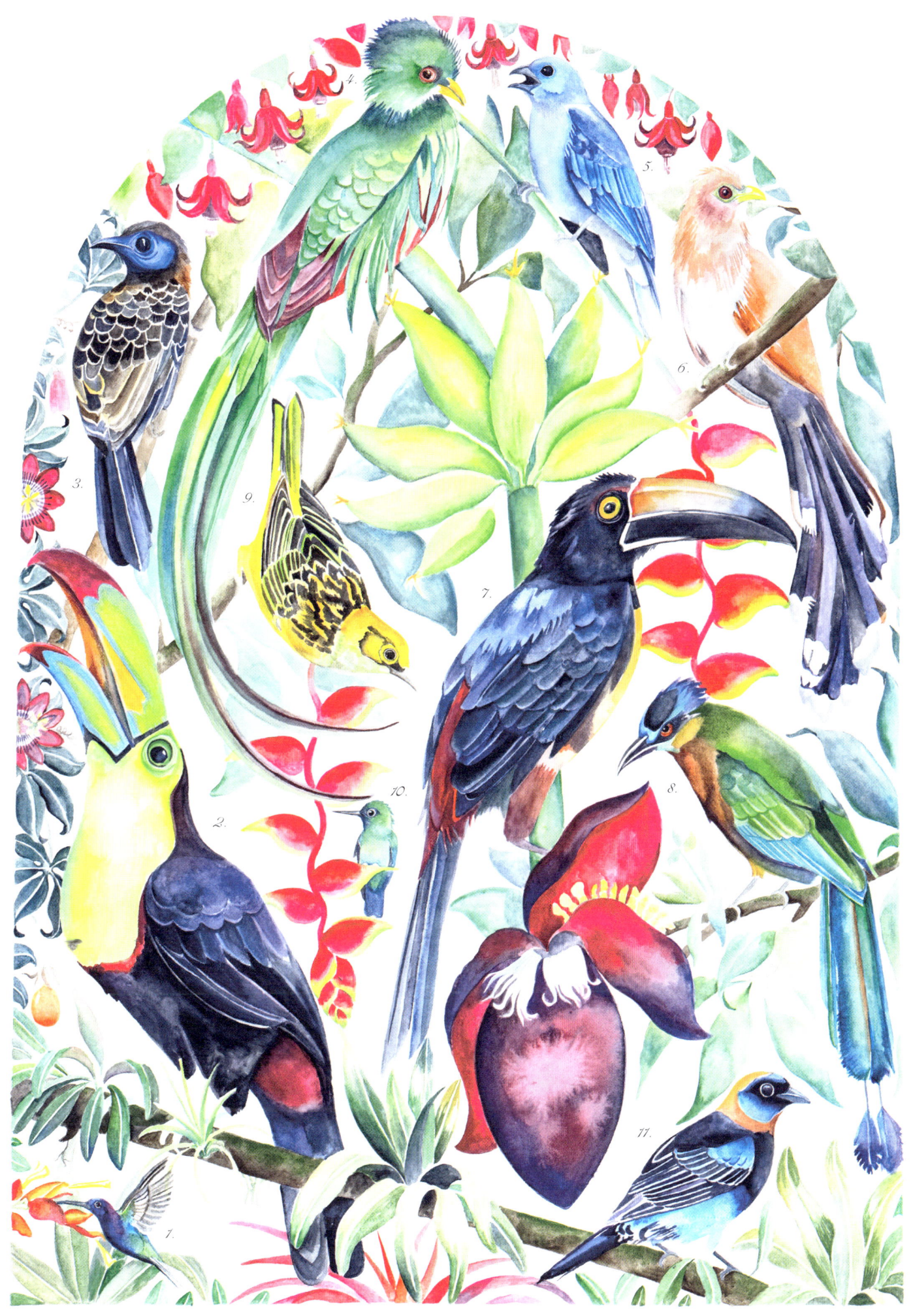

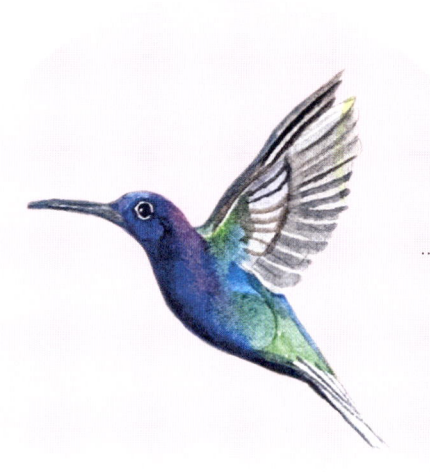

VIOLET SABREWING

Campylopterus hemileucurus
Order Apodiformes

Measuring up to 6 inches in length, this species
is the largest hummingbird in Central America.

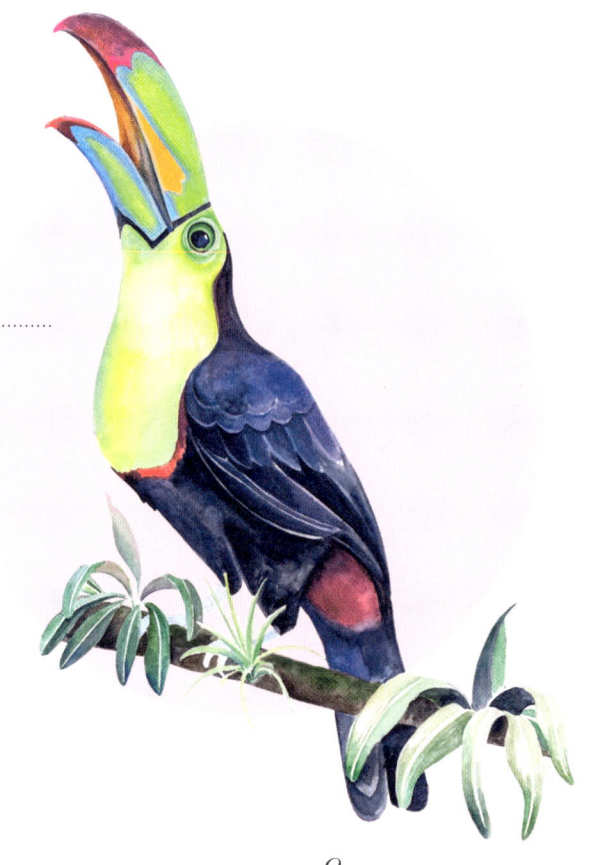

2.

KEEL-BILLED TOUCAN

Ramphastos sulfuratus
Order Piciformes

The toucan's large bill may be an evolutionary
anachronism that helped it eat particular fruits or
invertebrates. Scientists also determined that the
bill helps the bird stay cool on hot days, acting as a
thermal regulator that conserves or loses heat.

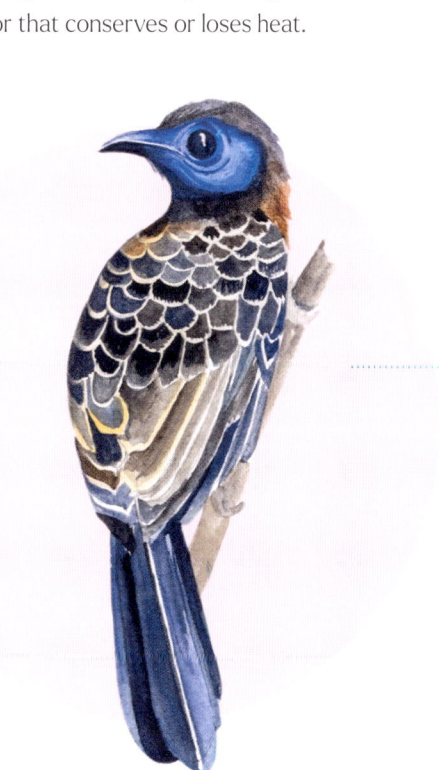

3.

OCELLATED ANTBIRD

Phaenostictus mcleannani
Order Passeriformes

Although these birds follow swarms of army ants, they
do not actually feed on the ants themselves but rather
on small insects flushed out by the ants' movement.

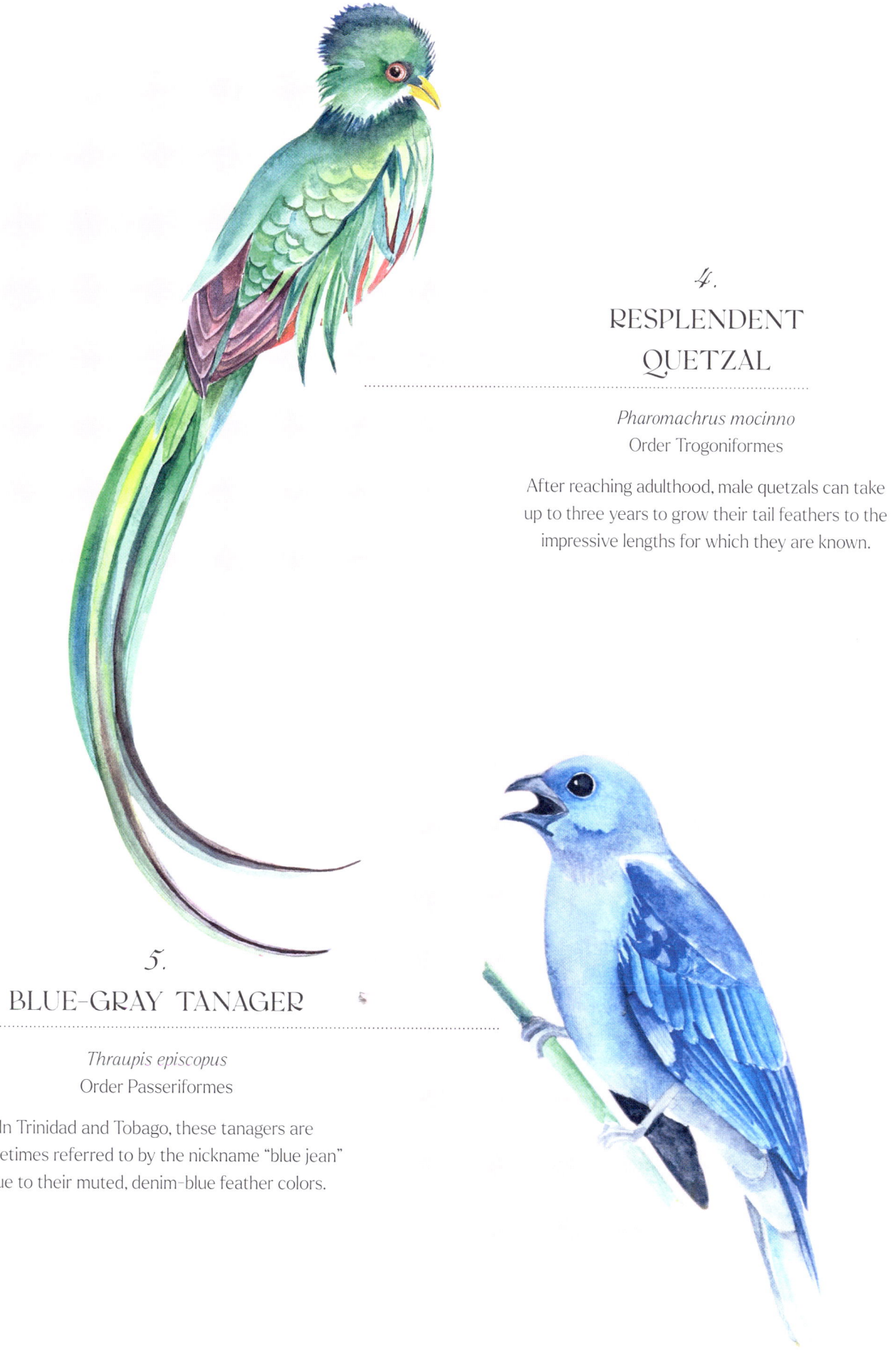

4.
RESPLENDENT QUETZAL

Pharomachrus mocinno
Order Trogoniformes

After reaching adulthood, male quetzals can take up to three years to grow their tail feathers to the impressive lengths for which they are known.

5.
BLUE-GRAY TANAGER

Thraupis episcopus
Order Passeriformes

In Trinidad and Tobago, these tanagers are sometimes referred to by the nickname "blue jean" due to their muted, denim-blue feather colors.

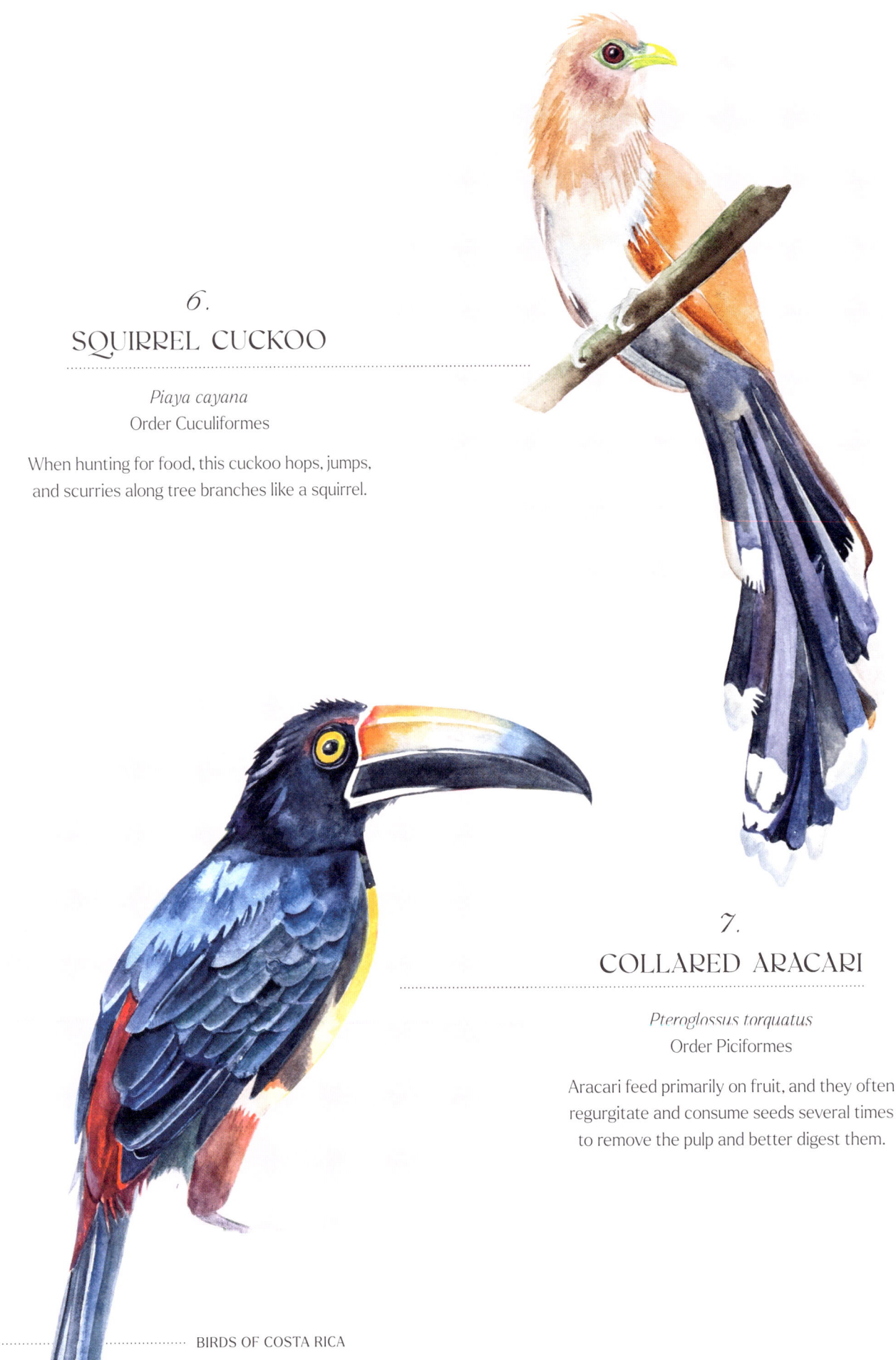

6.
SQUIRREL CUCKOO

Piaya cayana
Order Cuculiformes

When hunting for food, this cuckoo hops, jumps,
and scurries along tree branches like a squirrel.

7.
COLLARED ARACARI

Pteroglossus torquatus
Order Piciformes

Aracari feed primarily on fruit, and they often
regurgitate and consume seeds several times
to remove the pulp and better digest them.

8.
BLUE-CAPPED MOTMOT

Momotus coeruliceps
Order Coraciiformes

The word *motmot* comes from its distinctive call, which
sounds a bit like the double hoot of an owl.

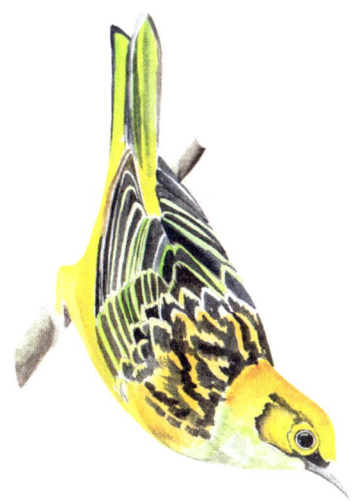

9.
SILVER-THROATED TANAGER

Tangara icterocephala
Order Passeriformes

These omnivores feed primarily on fruits and berries,
but they occasionally supplement their diet with
arthropods, especially during breeding season.

10.
COPPERY-HEADED EMERALD

Microchera cupreiceps
Order Apodiformes

This tiny hummingbird is endemic to Costa Rica, where it forages
on the nectar from a variety of flowers in the forest.

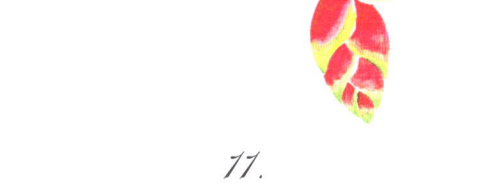

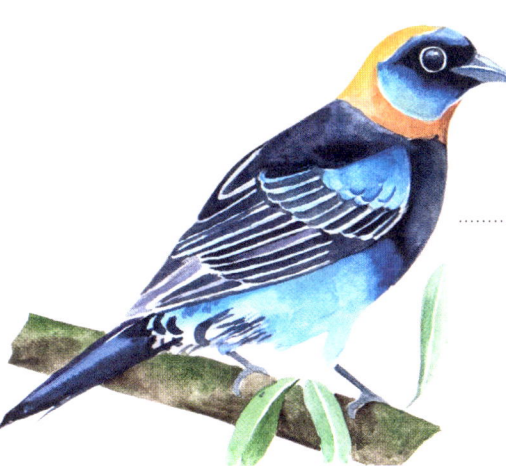

11.
GOLDEN-HOODED TANAGER

Stilpnia larvata
Order Passeriformes

In Costa Rica, this bird is often called *siete colores* (Spanish
for "seven colors") because of its multiple feather colors.

COASTAL BIRDS of CUBA and the CARIBBEAN

You are likely to meet several of these exceptional coastal birds as you're strolling along the white sand beaches or passing through snarled mangrove forests throughout the island nations in the Caribbean Sea. In fact, it would be challenging not to encounter a Brown Pelican (1). The national bird of several Caribbean nations, including Saint Martin and Barbados, the Brown Pelican is a well-known and common visitor. These ubiquitous, large, and unmistakable birds dive into the water and emerge with their large throat pouches filled with fish and small crustaceans. Their far-reaching range extends up the Pacific coast of North America. In fact, Alcatraz Island, a small island in the San Francisco Bay, was named by an eighteenth-century Spanish explorer who noted a vast number of the birds on the island (*alcatraz* is an ancient Spanish word for "pelican").

Another bird you are likely to spot soaring above the coastal waters of Cuba, easily identifiable by its unmistakable bulging red breast, is the Magnificent Frigatebird (3). The avian pirates of the Caribbean, these swashbuckling birds are known to swoop in and steal food from the mouths of other birds in midair. This habit has earned frigates their other name, "man-o-war birds," as they are known to harass other birds until they can steal their catch for themselves.

If you have a phobia of snakes, try not to be alarmed at the sight of the Anhinga (4), or snake-bird, which has a long, curved, slender neck and a snakelike head that pokes above the surface of the water as it swims. Its name is derived from a word in the Brazilian Tupi language and translates to "devil bird." Like the lookalike cormorants, Anhingas can often be seen sunbathing with wings outstretched to dry their waterlogged feathers.

1.
BROWN PELICAN

2.
YELLOW-HEADED
WARBLER

3.
MAGNIFICENT
FRIGATEBIRD

4.
ANHINGA

5.
NORTHERN JACANA

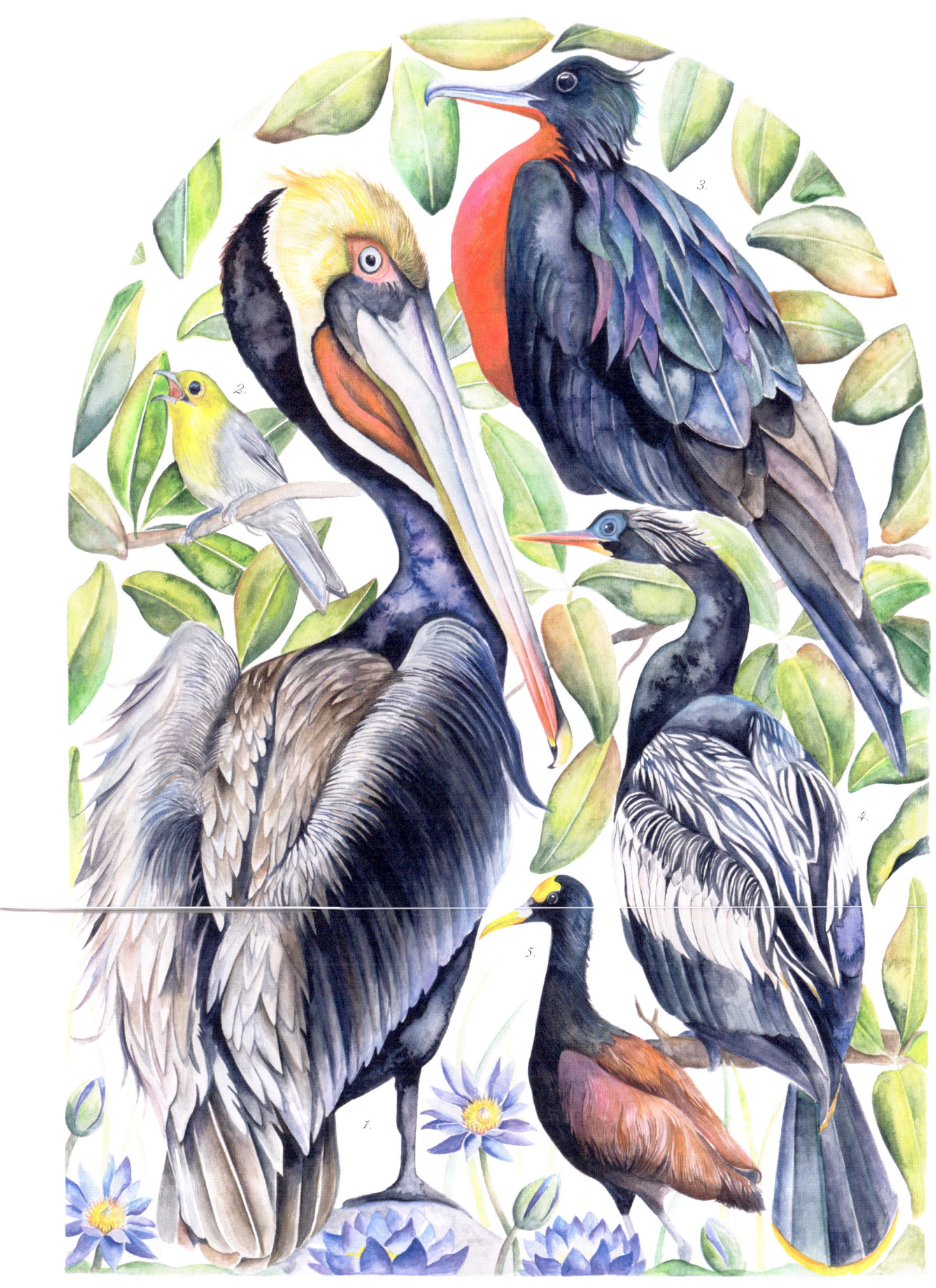

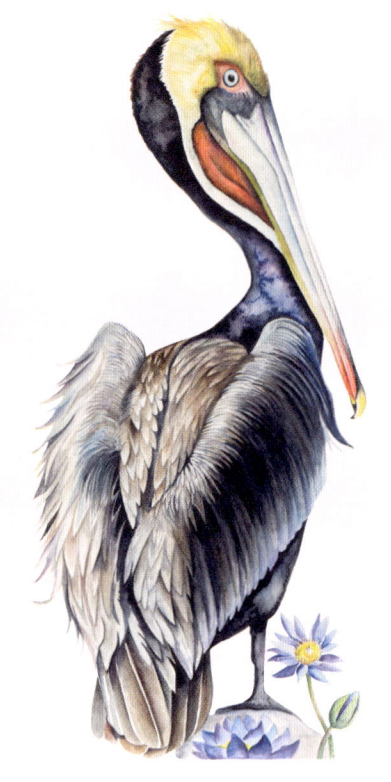

1.

BROWN PELICAN

Pelecanus occidentalis
Order Pelecaniformes

With their large, pouched bills, Brown Pelicans
are so successful at fishing that seagulls
sometimes steal fish right out of their mouths.

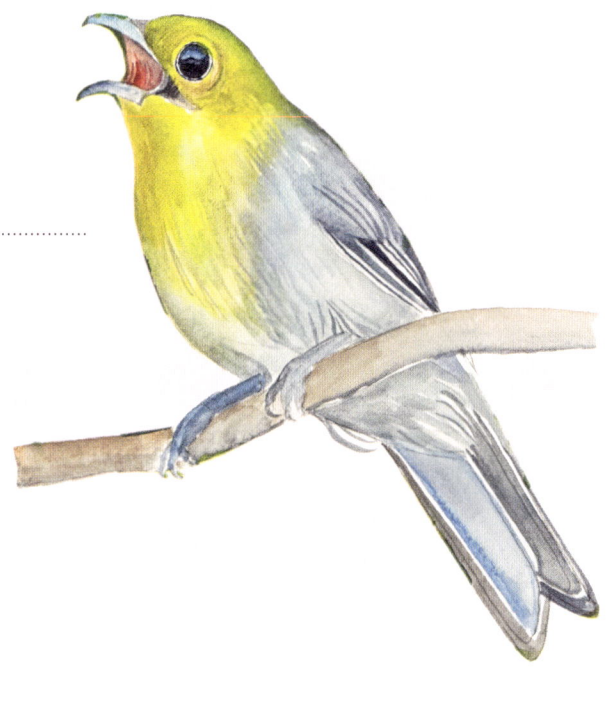

2.

YELLOW-HEADED WARBLER

Teretistris fernandinae
Order Passeriformes

This small, yellow-and-gray warbler is endemic to
western Cuba, where it forages from the ground
for insects and other small invertebrates.

3.

MAGNIFICENT FRIGATEBIRD

Fregata magnificens
Order Suliformes

Male Magnificent Frigatebirds have a striking
bright red breast pouch that they inflate to
attract females. Females have a white chest
and are much less distinctive looking.

4.
ANHINGA

Anhinga anhinga
Order Suliformes

These waterbirds do not have waterproof feathers. When damp, their feathers are heavy, which helps the birds submerge below the surface to fish. They are often seen drying their outstretched wings in the sun.

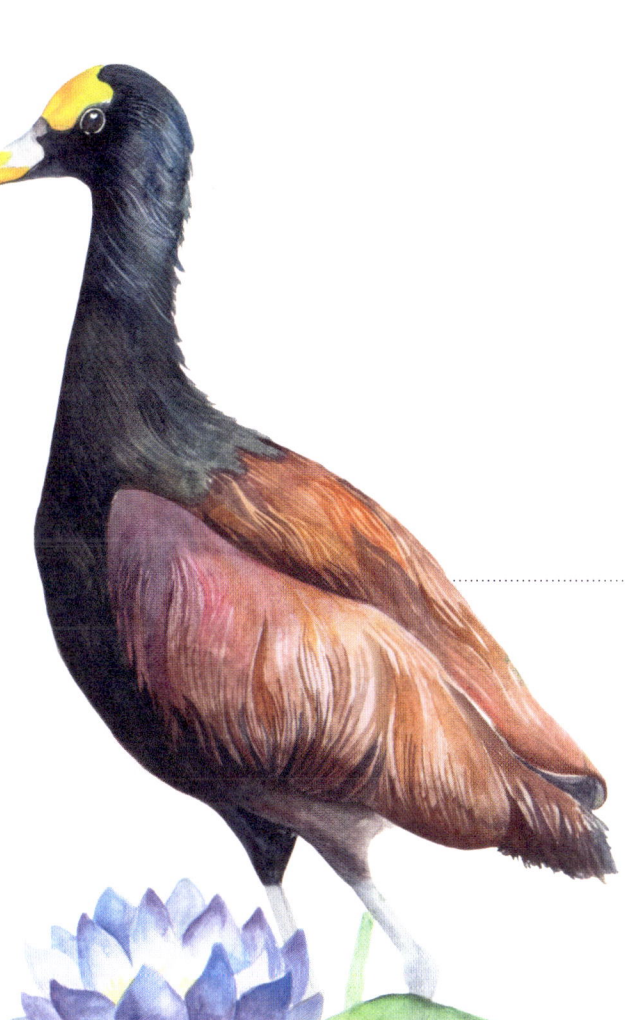

5.
NORTHERN JACANA

Jacana spinosa
Order Charadriiformes

This bird's extremely large feet enable it to balance and walk on floating vegetation, so it appears to be walking on the water's surface.

BIRDS
of the
FALKLAND
ISLANDS

This archipelago in the farthest southern reaches of the Atlantic Ocean is located at the border between the subantarctic oceanic and tundra climate zones. Despite the islands' forbidding and hostile climate, they have always been home to an abundance of birds.

Given that the islands are almost devoid of trees, birds rely on shrubs, rocks, and vegetation that have adapted to withstand the strong winds that frequently billow across the vast land. The Long-tailed Meadowlark (4), for example, prefers open, grassy habitats and forages on the ground, making it well suited to the treeless territory. Though some of the Falklands' bird species are also found in nearby southern Argentina and farther south in Antarctica, a few, including the Cobb's Wren (7) and a subspecies of the Long-tailed Meadowlark, are endemic to the islands.

The most iconic island residents are the penguins, including the Southern Rockhopper Penguin (8), a small, gregarious penguin with a yellow crest that nests on the rocky shores. Though many penguins travel by sliding on their bellies, Southern Rockhoppers hop from rock to rock and across the cracks and craggy shorelines. Their unique coloration provides useful camouflage while swimming in open water. When viewed from above, the penguin's black back blends into the reflected dark surface of the ocean; from below, its bright white belly blends in with the bright light of the sun cast onto the water from above.

In the eighteenth century at the height of the whaling industry, the number of penguins in the Falklands dipped significantly, as whaling ships visited the islands and crew members hunted the birds. As whaling began to decline in the nineteenth century, the penguin population rebounded, but numbers have since declined following various environmental threats. Today, these penguins are classified as an endangered species.

1.	*3.*	*5.*	*7.*
TWO-BANDED PLOVER	MAGELLANIC CORMORANT	BROWN SKUA	COBB'S WREN
2.	*4.*	*6.*	*8.*
BLACKISH OYSTERCATCHER	LONG-TAILED MEADOWLARK	BLACK-CHINNED SISKIN	SOUTHERN ROCKHOPPER PENGUIN

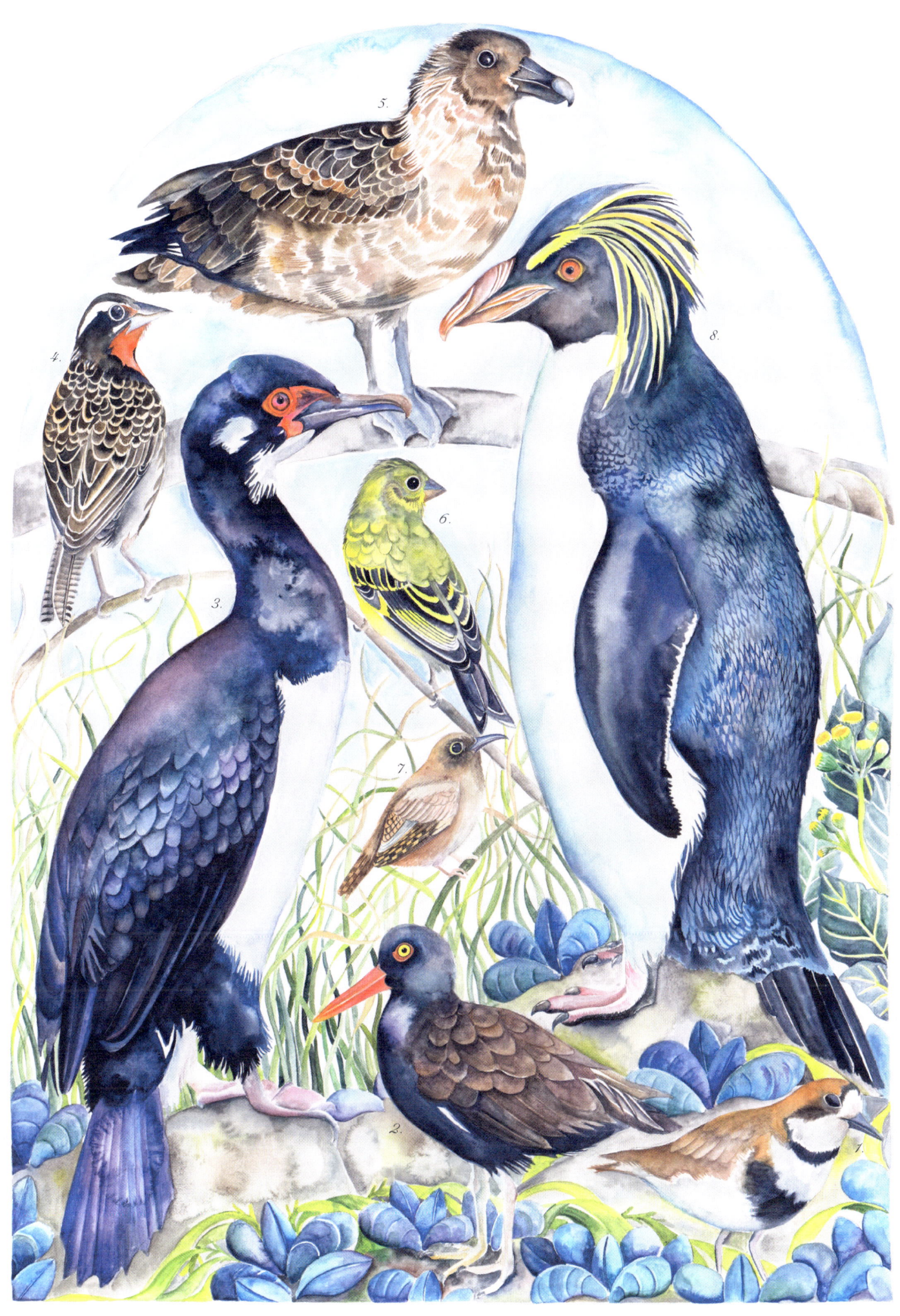

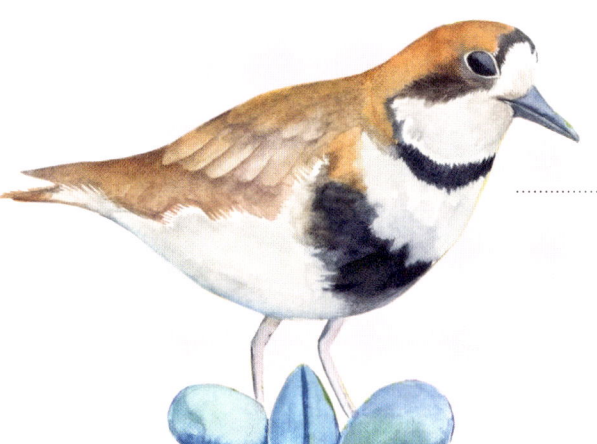

1.
TWO-BANDED PLOVER

Anarhynchus falklandicus
Order Charadriiformes

Outside the breeding season, these social birds gather in large flocks, sometimes numbering in the hundreds.

2.
BLACKISH OYSTERCATCHER

Haematopus ater
Order Charadriiformes

Contrary to what its common name suggests, the Blackish Oystercatcher eats mostly mussels and limpets, along with a variety of other mollusks—including oysters.

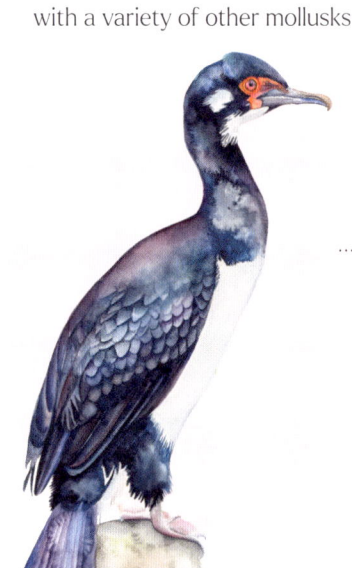

3.
MAGELLANIC CORMORANT

Leucocarbo magellanicus
Order Suliformes

Sharing a geographic range with the larger Imperial Cormorants, these birds tend to fish relatively close to shore, while the Imperial is known for feeding farther out in deep, open water.

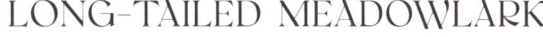

4.
LONG-TAILED MEADOWLARK

Leistes loyca falklandicus
Order Passeriformes

This endemic subspecies, found only in the Falkland Islands, is sometimes locally referred to as a "military starling," perhaps because these birds aggressively defend their territory.

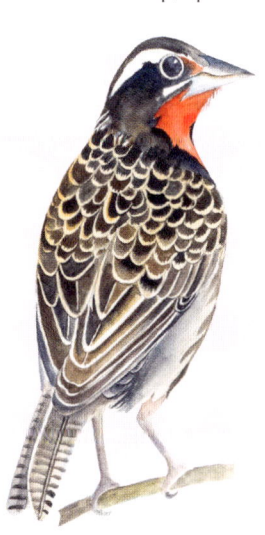

5.
BROWN SKUA

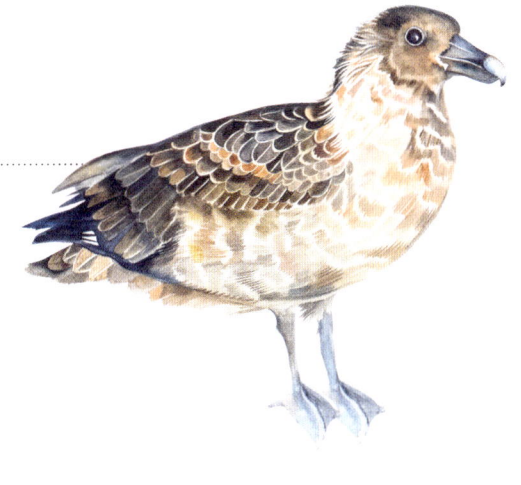

Stercorarius antarcticus
Order Charadriiformes

Some studies have suggested that these highly intelligent
birds can recognize individual human faces.

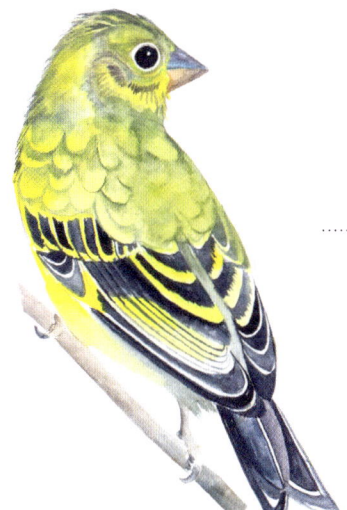

6.
BLACK-CHINNED SISKIN

Spinus barbatus
Order Passeriformes

This little finch's range extends farther south
than that of any other siskin.

7.
COBB'S WREN

Troglodytes cobbi
Order Passeriformes

Long considered to be a subspecies of the more common
House Wren, in 1993 the Cobb's Wren was confirmed to
be a unique species endemic to the Falkland Islands.

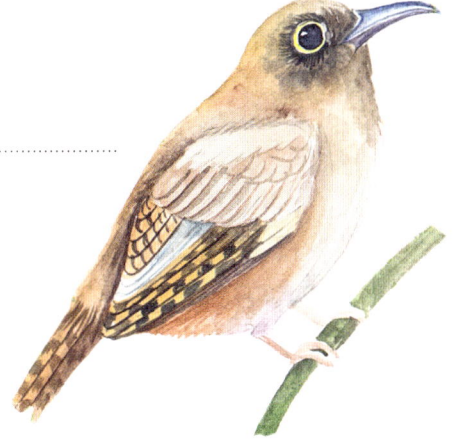

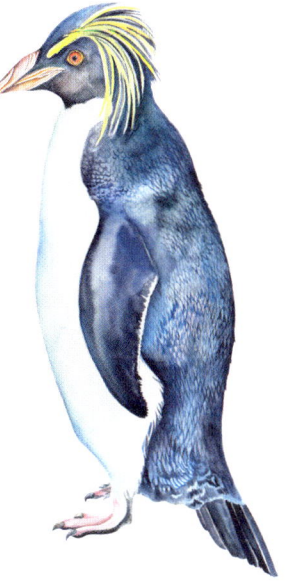

8.
SOUTHERN ROCKHOPPER PENGUIN

Eudyptes chrysocome
Order Sphenisciformes

These little penguins have a second transparent eyelid that serves
as a goggle-like barrier and enables them to see underwater.

BIRDS
of the
ATLANTIC
FOREST

The Atlantic Forest, which wraps along Brazil's Atlantic coast and extends inland into Paraguay, may be most widely known for beasts like the jaguar and the three-toed sloth, but it is also teeming with a truly spectacular array of birds.

South America is home to several species of guans, a group of birds that are found almost exclusively on the continent and have no close relatives. The Black-fronted Piping Guan (1), a large bird of the Atlantic Forest, vaguely resembles a pheasant. Unfortunately, deforestation has had a devastating impact on populations of these birds, which are now considered endangered.

The brilliant yellow, curve-billed Saffron Toucanet (5) is one of several birds in the toucan family that call this forest home. In Portuguese, it is known as *araçari-banana*, or "banana toucanet," for its vivid yellow plumage. This bird's distinctive yellow color has led to its illegal hunting and capture to sell in the caged bird industry. Saffron Toucanet populations, which have already been threatened by significant habitat loss, have correspondingly suffered immensely as a result.

The intrepid birdwatcher may come across one of the scarcer and more difficult to find birds of the Atlantic Forest. The Atlantic Royal Flycatcher (7) reveals its colorful fanned crest only when trying to impress a mate. These birds are quite rare and therefore relatively little is known about them. Those who are lucky enough to encounter one will not soon forget it.

1.
BLACK-FRONTED
PIPING-GUAN

2.
DIADEMED TANAGER

3.
PIN-TAILED MANAKIN

4.
YELLOW-FRONTED
WOODPECKER

5.
SAFFRON TOUCANET

6.
MAROON-BELLIED
PARAKEET

7.
ATLANTIC ROYAL
FLYCATCHER

8.
SURUCUA TROGON

9.
GIANT ANTSHRIKE

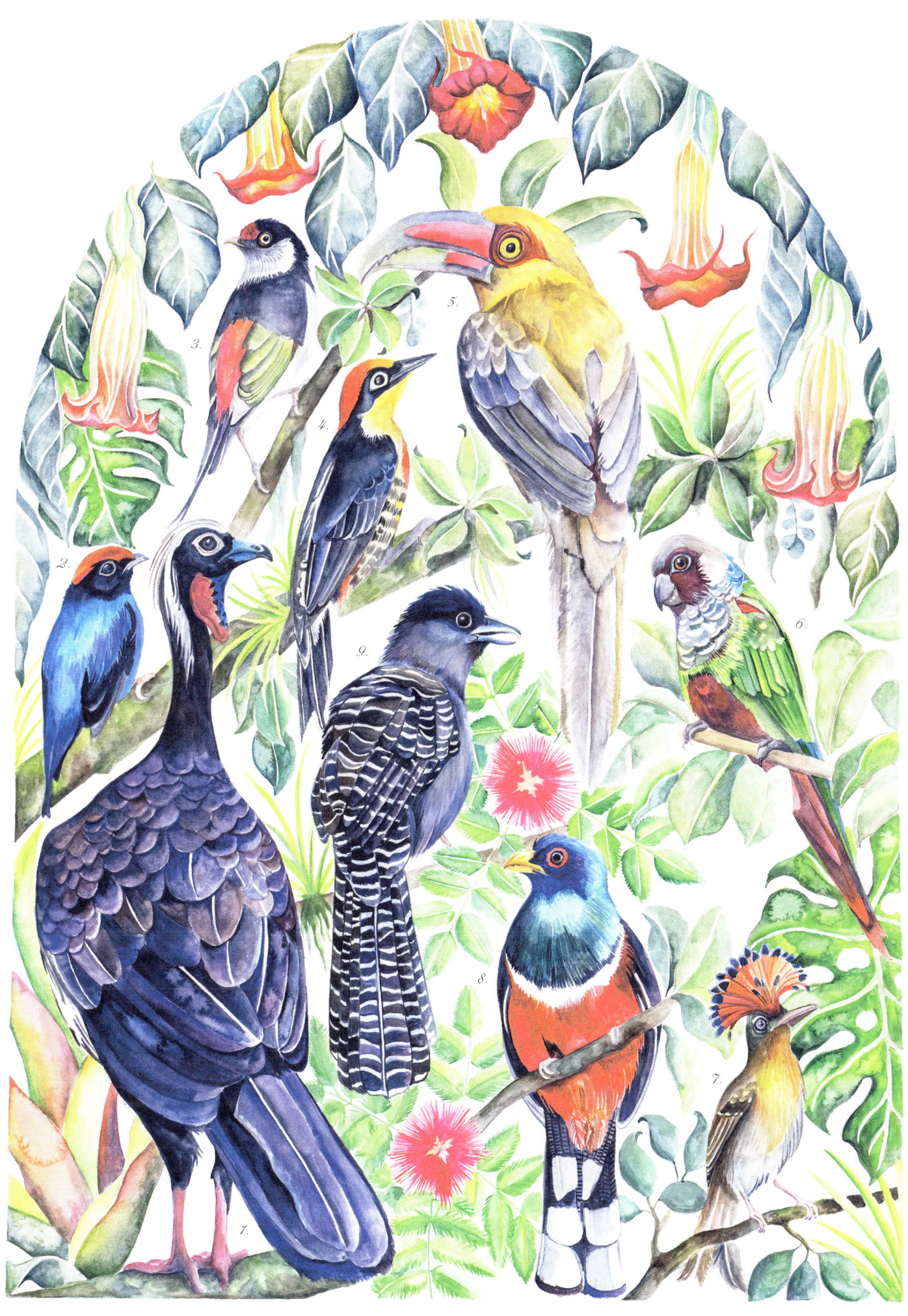

1.

BLACK-FRONTED
PIPING-GUAN

Pipile jacutinga
Order Galliformes

Since 2004, this beautiful bird has been
listed as an endangered species due to
hunting and severe habitat loss.

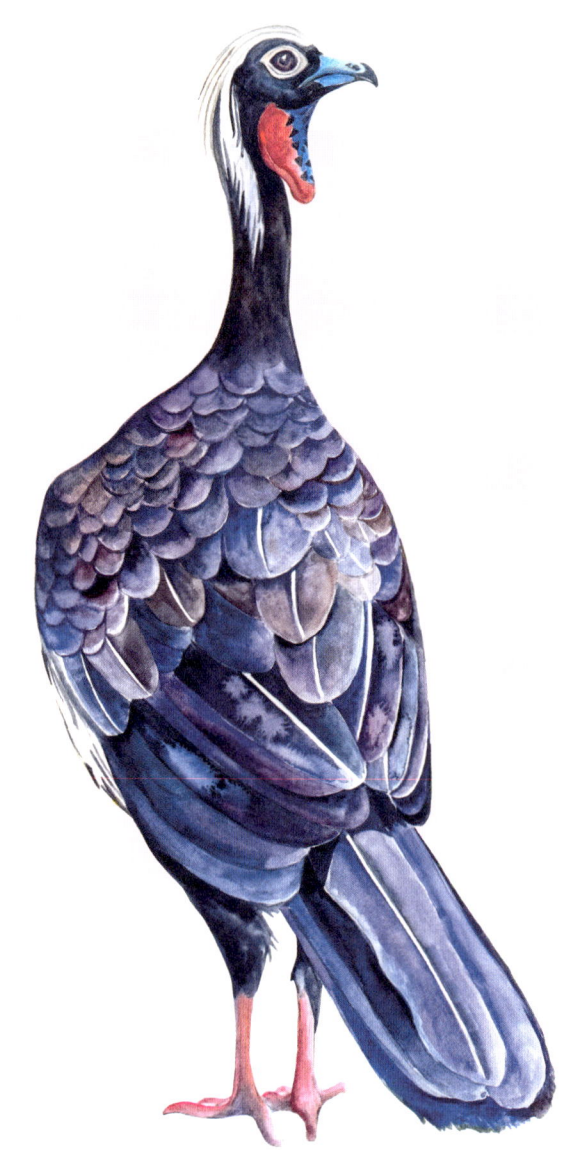

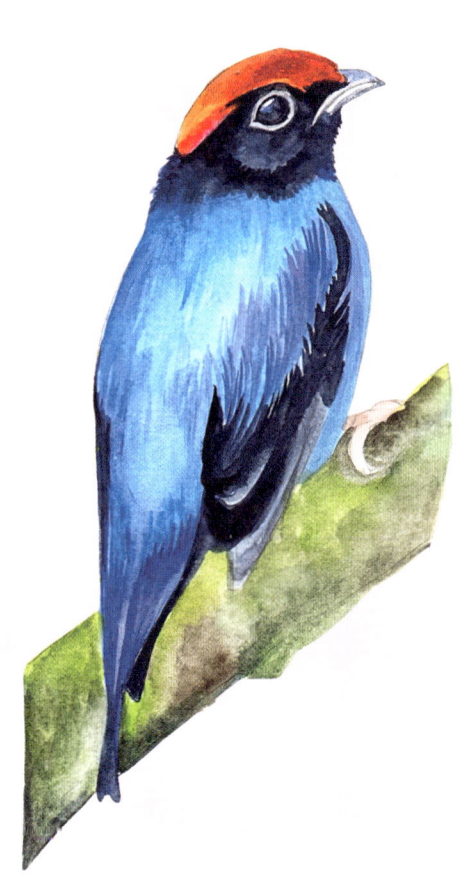

2.

DIADEMED TANAGER

Stephanophorus diadematus
Order Passeriformes

This purple-blue bird is the only member
of the genus *Stephanophorus*.

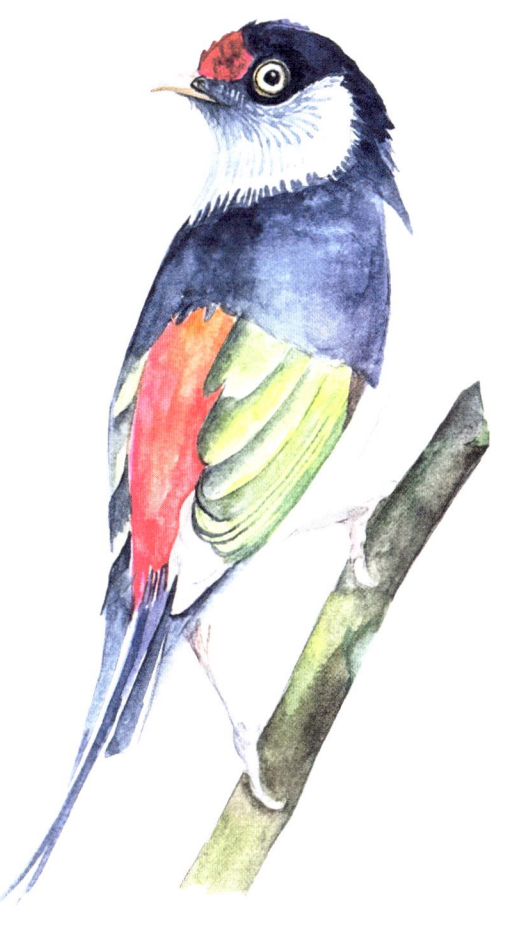

3.
PIN-TAILED MANAKIN

Ilicura militaris
Order Passeriformes

These colorful little birds derive their
common name from the Middle Dutch word
mannekijn, which means "little man."

4.
YELLOW-FRONTED
WOODPECKER

Melanerpes flavifrons
Order Picidae

These woodpeckers are cooperative breeders,
meaning several male and female birds may together
care for a single nest of eggs and hatchlings.

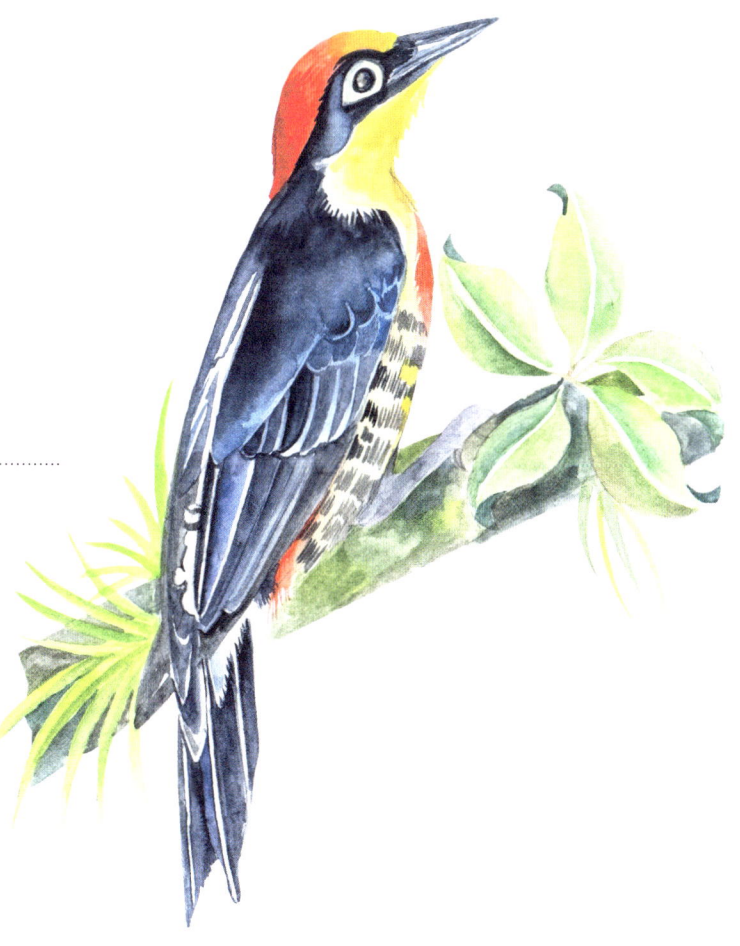

33

BIRDS OF THE ATLANTIC FOREST

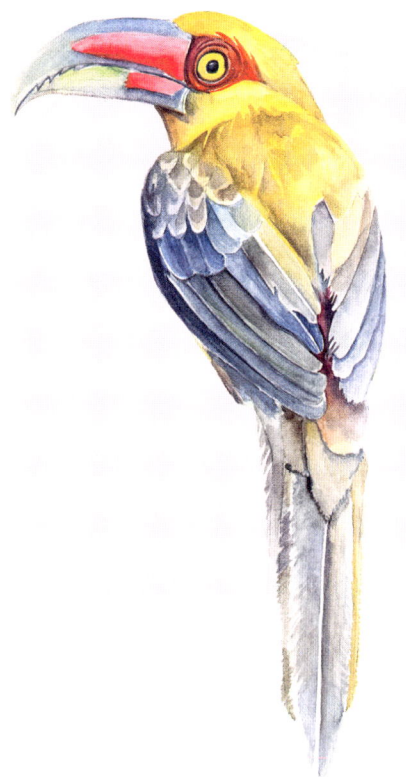

5.
SAFFRON TOUCANET

Pteroglossus bailloni
Order Piciformes

Although fruit makes up most of its diet, this saffron-yellow toucan will occasionally eat other birds' eggs and even smaller birds or their nestlings.

6.
MAROON-BELLIED PARAKEET

Pyrrhura frontalis
Order Psittaciformes

These parakeets can learn to talk, but their very shrill, unpleasant voices bother some people.

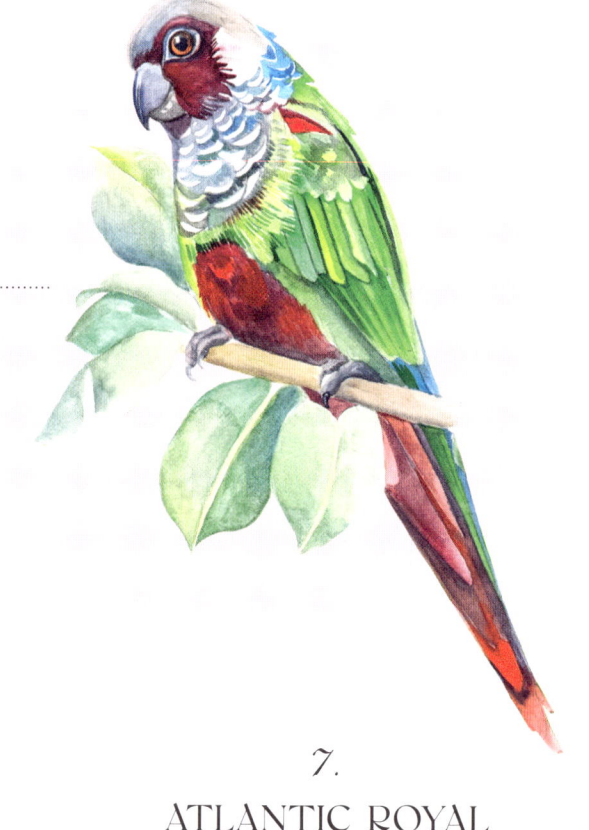

7.
ATLANTIC ROYAL FLYCATCHER

Onychorhynchus swainsoni
Order Passeriformes

These impressive hunters can snatch insects in midair using dizzying acrobatic techniques.

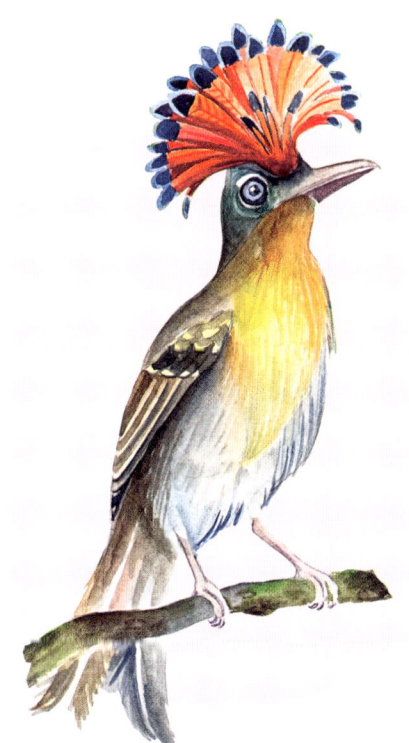

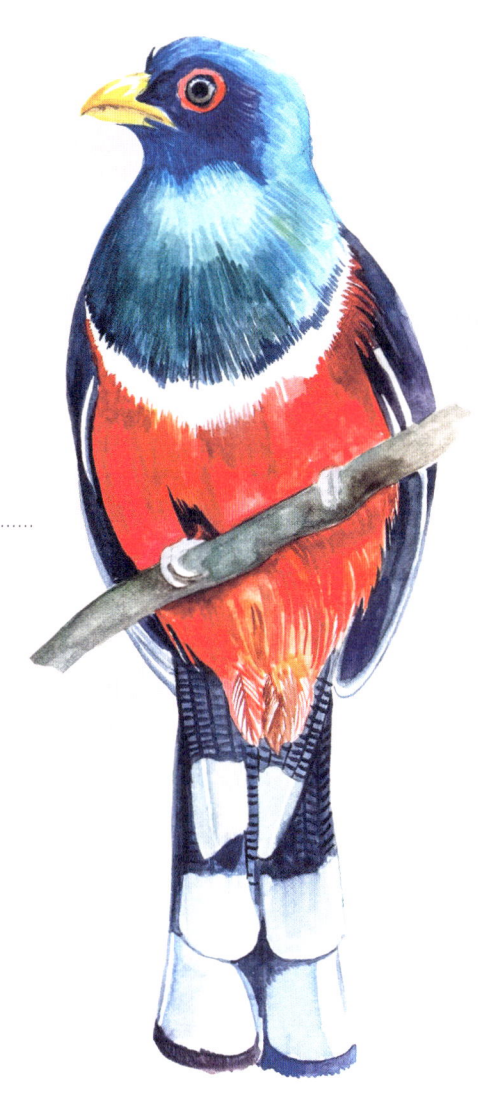

8.
SURUCUA TROGON

Trogon surrucura
Order Trogoniformes

The word *trogon* comes from the Greek word
for "nibbling" and refers to this bird's habit
of gnawing holes in trees to make nests.

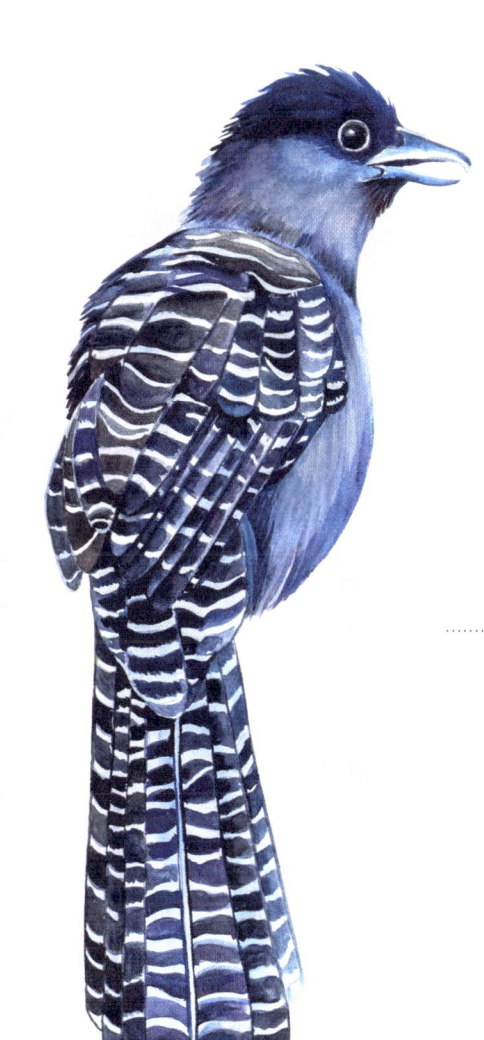

9.
GIANT ANTSHRIKE

Batara cinerea
Order Passeriformes

At up to 14 inches long, this is the
largest of all antbirds.

BIRDS
of the
GALÁPAGOS
ARCHIPELAGO

The Galápagos Islands, an archipelago of volcanic islands located far off the west coast of South America, is home to coastal, semi-desert, grassland, and dry forest habitats. The islands are renowned for their notable range of endemic species, many of which are birds. In the early nineteenth century, British naturalist Charles Darwin traveled to the remote islands to conduct ecological research. His studies of Galápagos birds provided much of the inspiration for what would become his theory of evolution.

Easily identified by its electric-blue feet, the Blue-footed Booby (3) must be one of the most iconic birds of the Galápagos Islands. Its foot color is a result of the way it processes its diet of fish and indicates the relative health and diet of the individual bird. Birds with the bluest feet are considered the most desirable partners for mating.

Among the cacti in the dry lowland areas of the islands lives the "avian vampire" of the Galápagos. Although the diet of the Galápagos Mockingbird (4) consists mainly of insects, fruit, and small crabs, these birds have been known to drink the blood of wounded iguanas and seabirds that are still very much alive.

Another highly unusual bird that is endemic to the Galápagos is the Swallow-tailed Gull (5), which feeds almost entirely nocturnally—the only species of gull to do so. These gulls forage for fish and squid at night, hunting the creatures as they swim close enough to the surface of the water to be plucked out from above.

1.
LAVA HERON

2.
GALÁPAGOS DOVE

3.
BLUE-FOOTED
BOOBY

4.
GALÁPAGOS
MOCKINGBIRD

5.
SWALLOW-TAILED
GULL

6.
RED-FOOTED
BOOBY

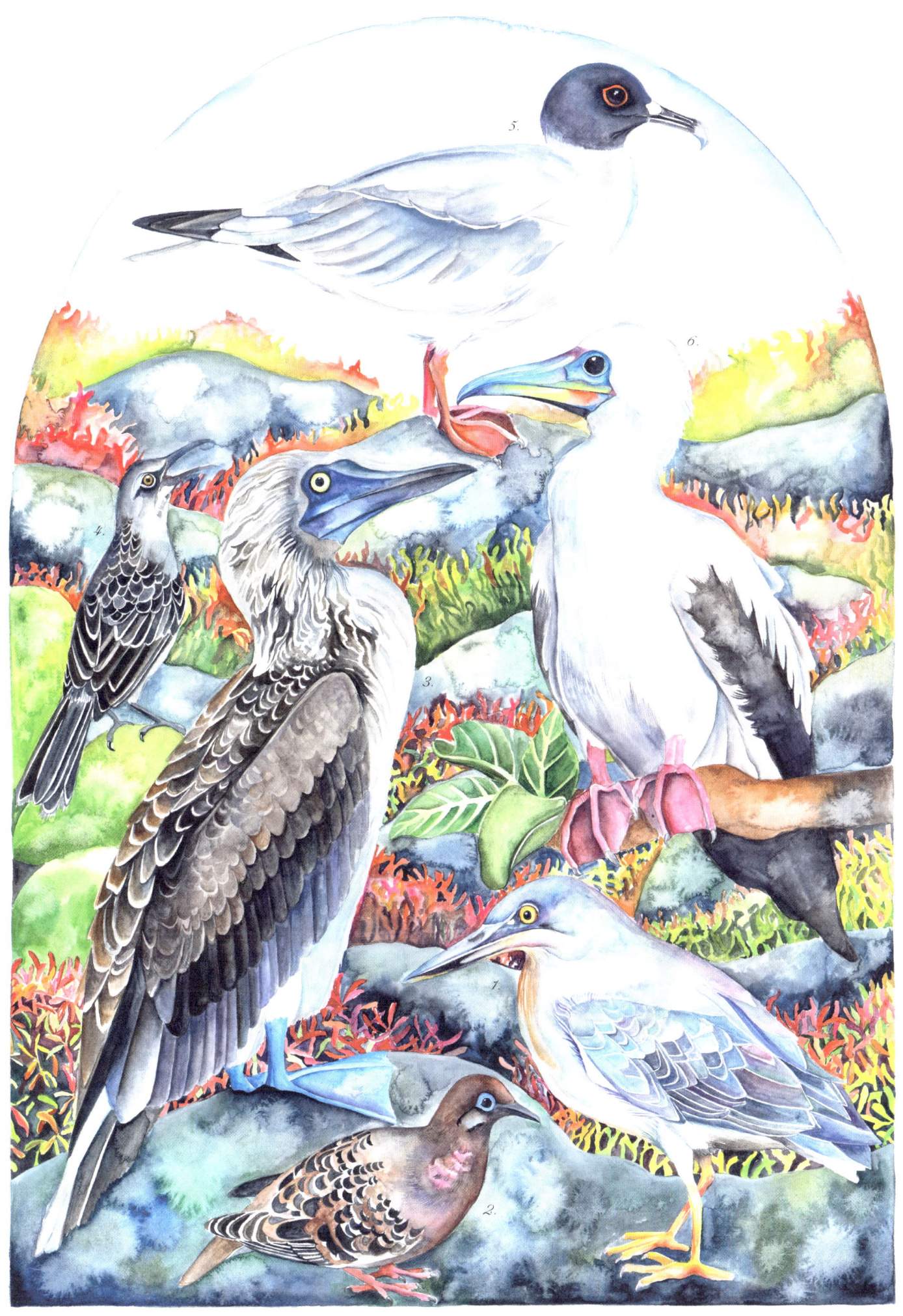

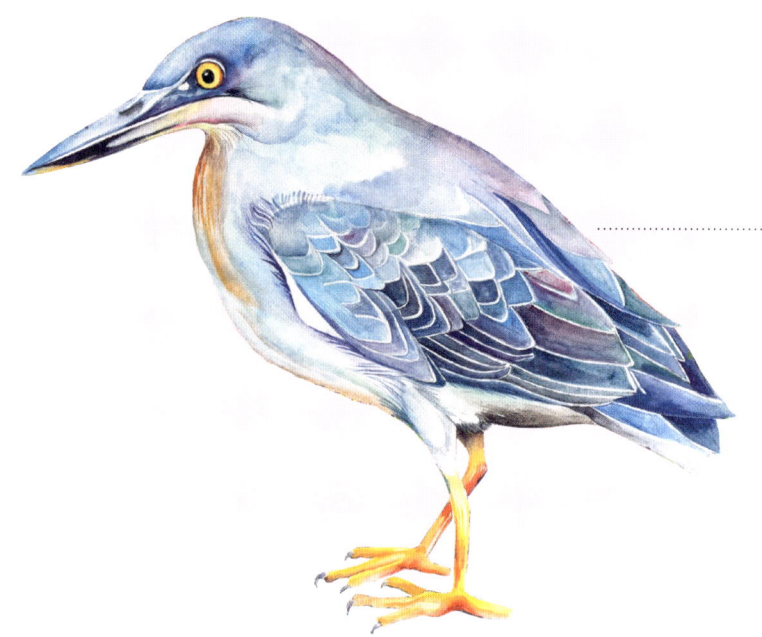

1.
LAVA HERON

Butorides sundevalli
Order Pelicaniformes

This heron's gray-blue coloring enables it to blend in along the shoreline, where it hunts for crabs and fish and spears its prey with its long, sharp beak.

2.
GALÁPAGOS DOVE

Zenaida galapagoensis
Order Columbiformes

When this dove believes its nest is threatened, it will feign injury to lure a predator away from its eggs and nestlings.

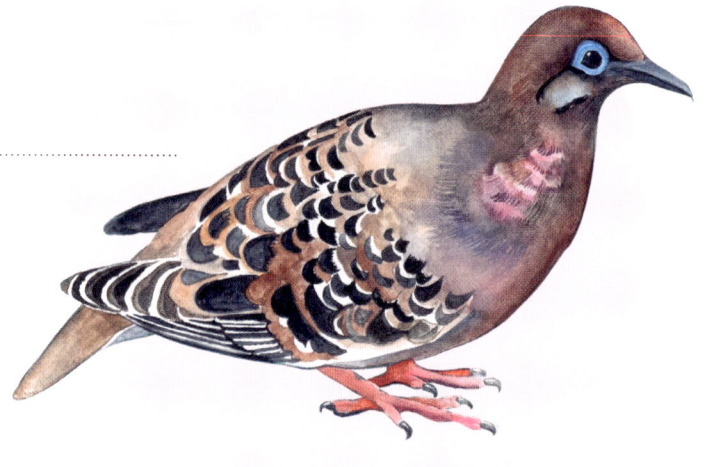

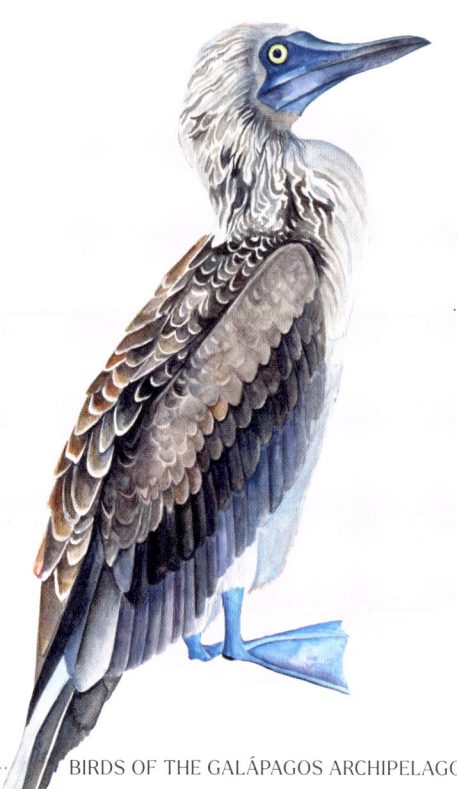

3.
BLUE-FOOTED BOOBY

Sula nebouxii
Order Suliformes

Instead of building a nest, this booby lays its eggs directly on the ground. It defecates while incubating to create a circular wall of excrement surrounding the eggs.

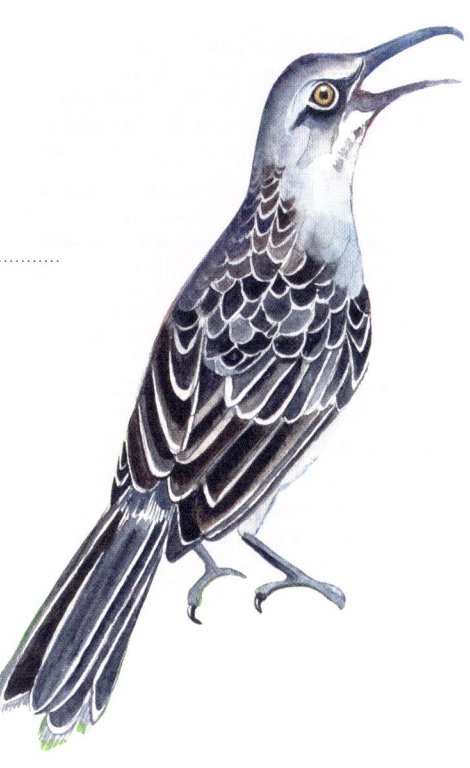

4.
GALÁPAGOS MOCKINGBIRD

Mimus parvulus
Order Passeriformes

These highly sociable birds are cooperative breeders that form family groups of up to twenty birds that collaborate to raise young.

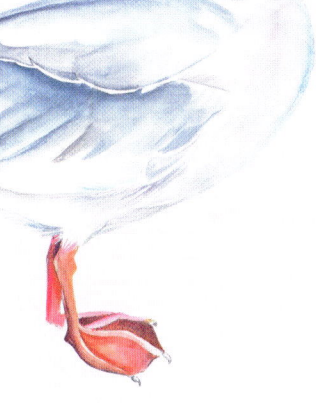

5.
SWALLOW-TAILED GULL

Creagrus furcatus
Order Charadriiformes

When danger threatens their colony, these gulls make loud vocalizations to warn others of the coming threat.

6.
RED-FOOTED BOOBY

Sula sula
Order Suliformes

These seabirds prefer nesting sites that are in full, direct sun and generally in trees and bushes, unlike other boobies, which nest on the ground.

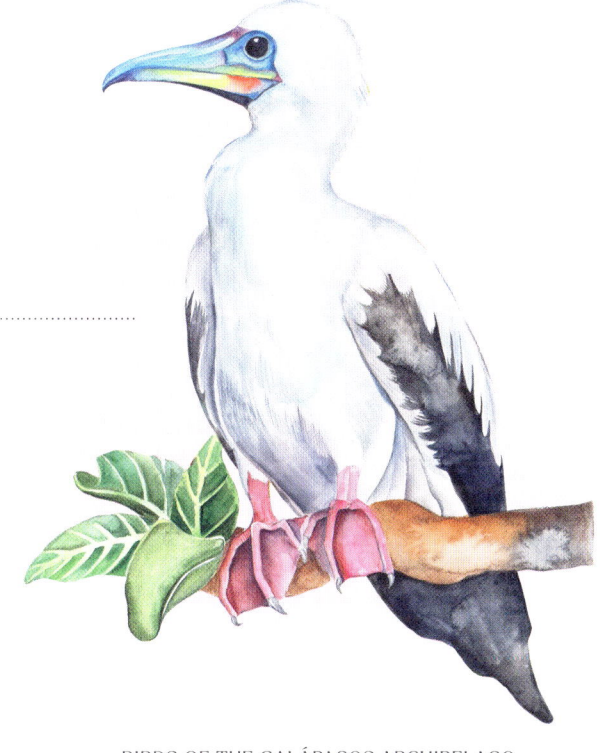

BIRDS
of the
AMAZON

The Amazon Rainforest is one of the most remarkable and wild expanses of jungle wilderness in the world. Perhaps most famous for the large concentration of slithering deadly snakes, playful primates, and technicolor poisonous frogs, it is also positively bursting with beautiful birds. The density of the foliage in the humid tangles of rainforest vegetation can make it quite dark, and often the prismatic colors of these birds are on full display only when caught in a ray of sunlight slicing through the dense canopy.

The Amazon is home to one of the largest of all the New World vultures—the King Vulture (1). This majestic bird of prey can be nearly 2 feet tall. It was worshipped and revered by the Mayans, who sometimes depicted it as a god with the body of a winged man and the colorful head of the vulture, flying toward the heavens. Its regal name is often attributed to ancient Mayan legend, wherein the winged creature carries messages between humans and the gods.

Tanagers, a large family of birds that includes more than 200 species, are common visitors to the Amazon. The word *tanager* comes from *tangara*, from the language of the Indigenous Tupi people of the Amazon region. Two tanager species, the brilliantly red Brazilian Tanager (2) and the rainbow-feathered Red-necked Tanager (4), catch the eye as they flicker through the trees looking for tender fruits.

One of the most mysterious birds of the Amazon is the Hoatzin (7), a bizarre-looking species with a mohawk-like feather crest. Scientists have spent years trying to categorize this peculiar bird, eventually concluding that the Hoatzin has no close living relatives. This bird's unique digestive system uses gut bacteria to digest the plants it consumes. Because this process emits a powerful odor, the bird is sometimes locally referred to as the "stinkbird."

1.
KING VULTURE

2.
BRAZILIAN TANAGER

3.
PLUSH-CRESTED JAY

4.
RED-NECKED TANAGER

5.
NORTHERN EMERALD-TOUCANET

6.
GREEN OROPENDOLA

7.
HOATZIN

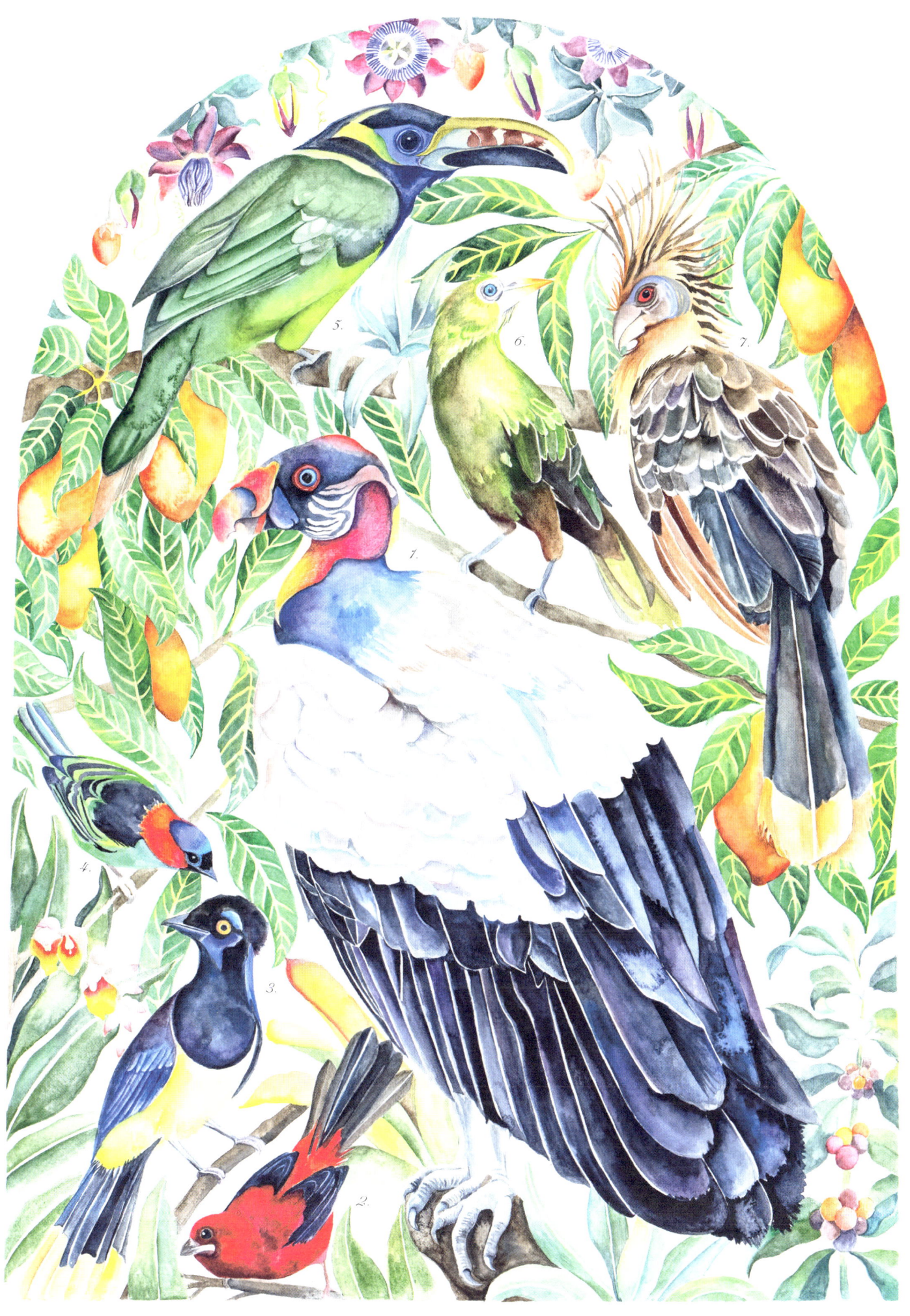

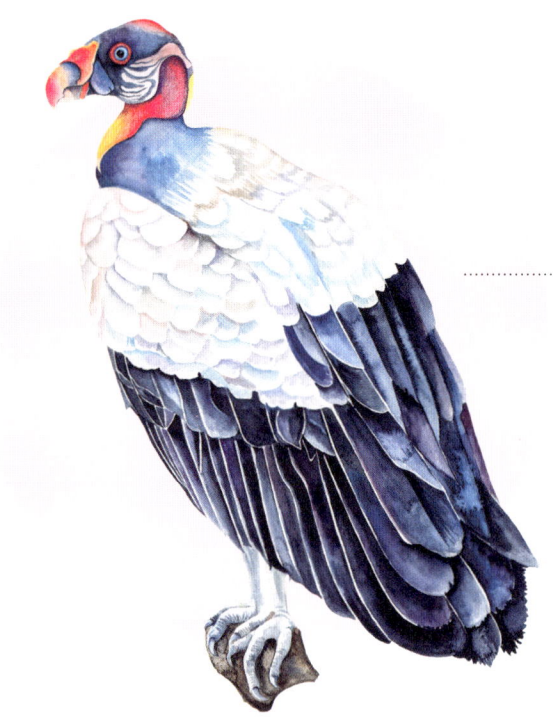

1.
KING VULTURE

Sarcoramphus papa
Order Cathartiformes

After building a nest directly on the ground, the female King Vulture lays a single egg. Then parents share the work of incubating the egg, with both taking turns on the nest.

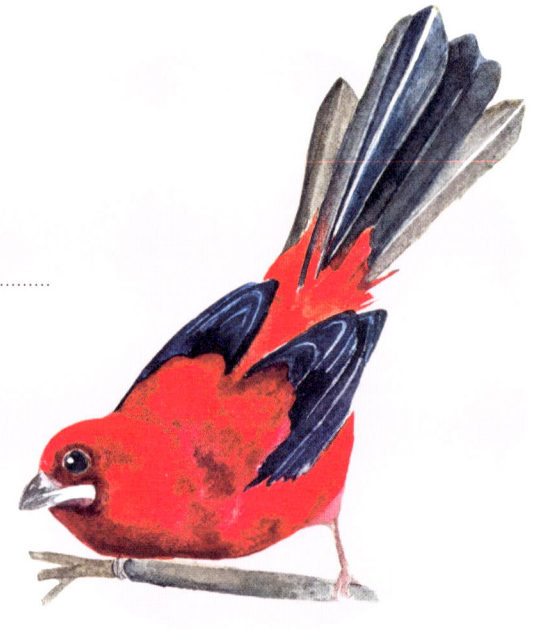

2.
BRAZILIAN TANAGER

Ramphocelus bresilius
Order Passeriformes

The brilliant red feathers of this Brazilian endemic bird are unique to the male of the species. The female is mostly brown and much less remarkable looking.

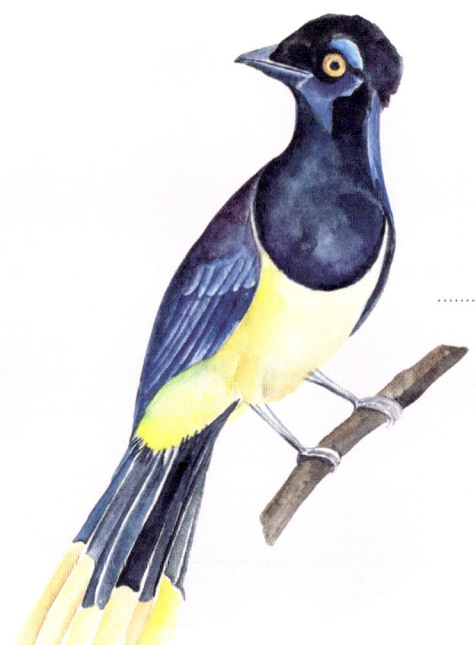

3.
PLUSH CRESTED JAY

Cyanocorax chrysops
Order Passeriformes

These birds are omnivores, and their diet includes fruit, insects, seeds, and sometimes small frogs and snakes.

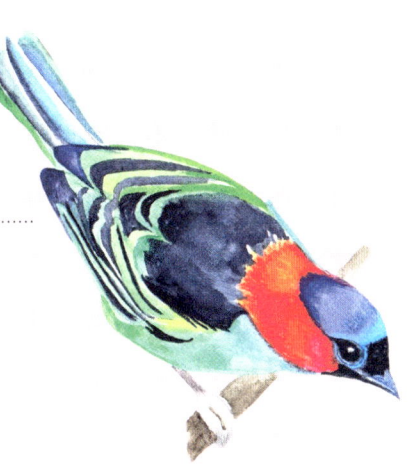

4.
RED-NECKED TANAGER

Tangara cyanocephala
Order Passeriformes

The juvenile Red-necked Tanager is mostly green and
lacks the brilliant red neck feathers of the adult.

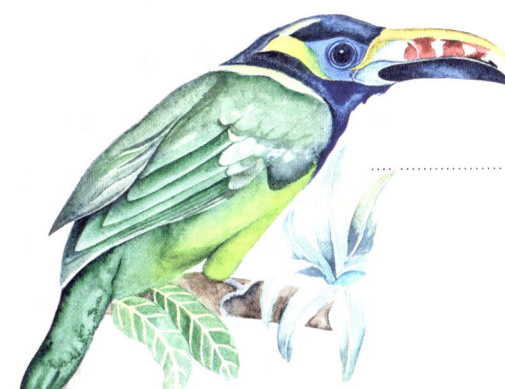

5.
NORTHERN EMERALD-TOUCANET

Aulacorhynchus prasinus
Order Piciformes

The patterning and coloration of this toucan, particularly on the
bill, varies slightly depending on the region in which it lives.

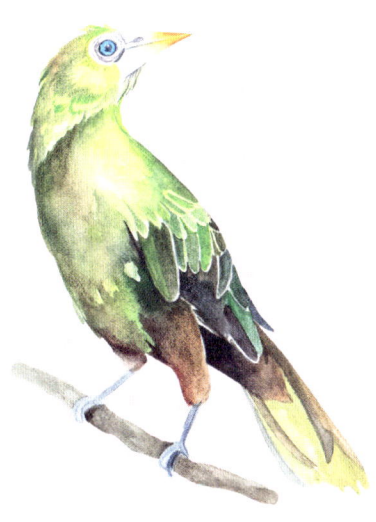

6.
GREEN OROPENDOLA

Psarocolius viridis
Order Passeriformes

These birds live in colonies and construct hanging
pendulous nests of woven plants and grasses.

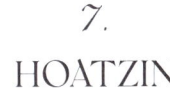

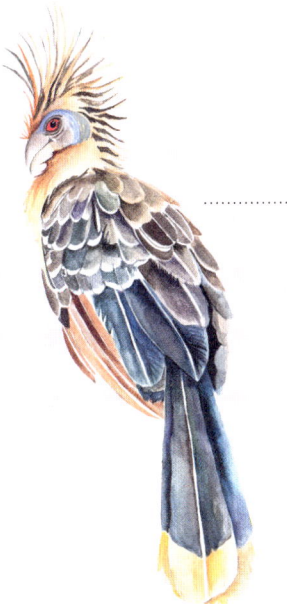

7.
HOATZIN

Opisthocomus hoazin
Order Opisthocomiformes

Hoatzin chicks hatch with two claws on each wing, which
help them move from branch to branch until they become
more coordinated. They lose their claws in maturity.

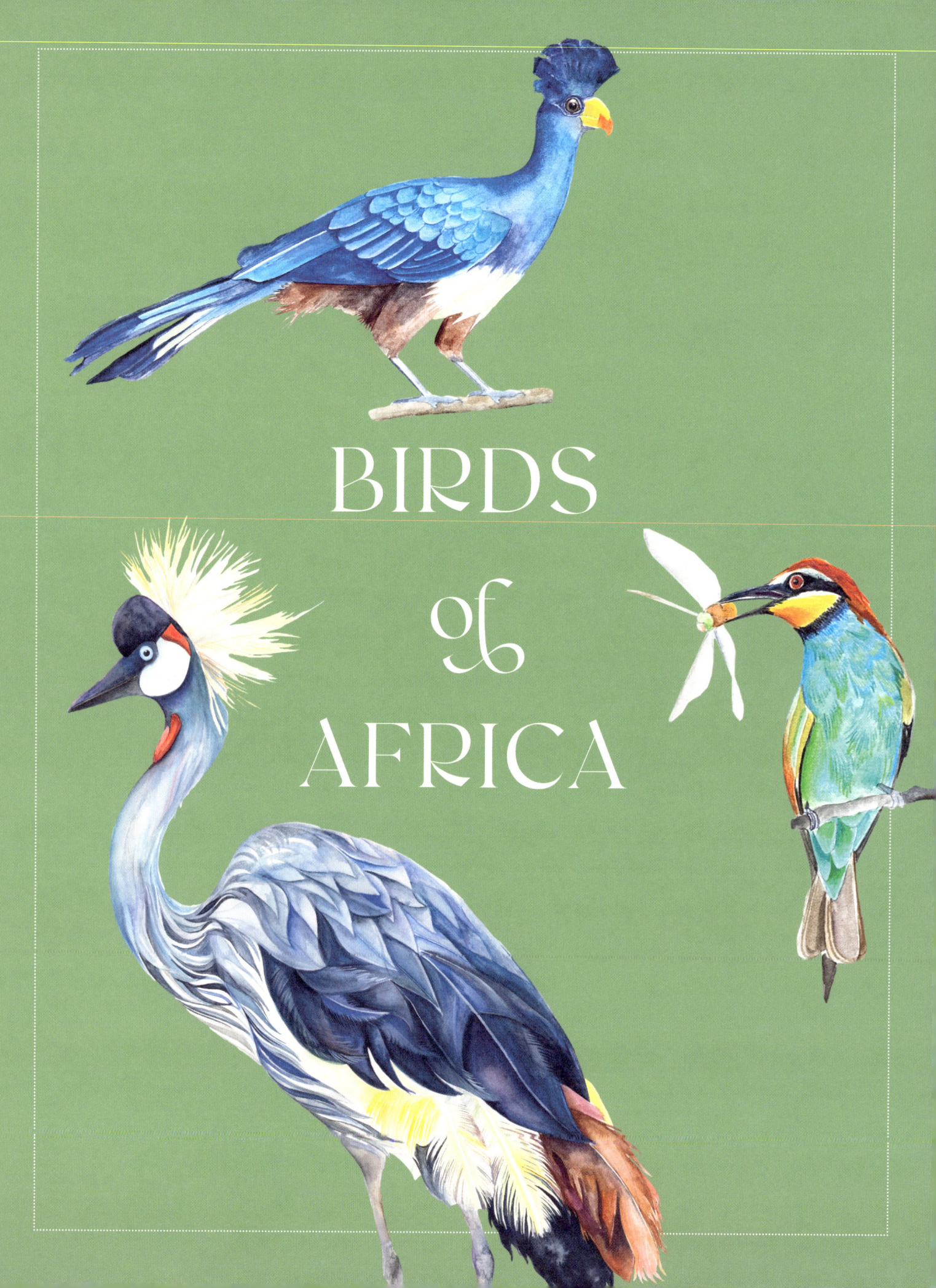

BIRDS

of

AFRICA

Although the "big five"—lions, elephants, rhinoceroses, buffalos, and leopards—might be the first animals that come to mind when you're thinking of African wildlife, the continent is also home to some of the most noteworthy birds in the world, which are not to be overshadowed by their mammalian counterparts. Many African cultures believe that certain birds provide links to ancestral spirits, and people therefore treat birds with tremendous respect and even reverence.

The Hoopoe, a bird prevalent throughout Africa, Asia, and Europe, has made its mark in cultures around the world. Look for the brilliant blue body and bizarre, ugly head of the Vulturine Guineafowl strutting through the grasslands, and scan the banks of the Nile in search of the ruddy-colored head of the Egyptian Goose.

BIRDS of SOUTHWESTERN AFRICA

Birds in Southern Africa hold a prominent place in the mythology and symbolism of various tribes and cultures in the region. Some birds are believed to be reincarnated human souls that have reached the highest level of perfection and taken flight. It is considered a great sin not only to harm a bird but also to harm a tree, as trees are the resting places of migratory birds. In Zulu, there is a saying: "The tree is given birth to by the bird." Birds rest in trees on their migration path and release excrement rich with the seeds of distant lands. Their presence in the trees symbolizes the growth of new trees and fertility and new life. In fact, in several African languages, the word for bird generally translates to mean "fat" or "fattening," which alludes to the belief that bird migrations at certain times of the year bring with them auspices of fertility, or a fattening of the land.

When birding in southwestern Africa, you may encounter the iridescent sheen and distinctive wrinkled pink head of the Southern Bald Ibis (1), or you may spy the brilliant architecture of a woven nest made by a Red-headed Weaver (4), which dangles pendulously over the banks of a river or along the fence of open farmland, like a delicate, braided sculpture. Perhaps you may be lucky enough to hear the distinctive call of the elusive Gorgeous Bushshrike (2).

Westerners would do well to use caution in using a dove or pigeon as a representative of peace in Africa. Contrary to many Western cultures, wherein doves and pigeons are symbols of peace and harmony, the Zulu people see these birds as symbols of war. This is because male pigeons and doves are known to fight to the death when vying for the attention of a female. This belief is echoed in the ancient beliefs of the Sumerians and Babylonians.

1.
SOUTHERN BALD IBIS

2.
GORGEOUS
BUSHSHRIKE

3.
AFRICAN PARADISE
FLYCATCHER

4.
RED-HEADED WEAVER

5.
GREAT BLUE TURACO

6.
VIOLET-EARED
WAXBILL

7.
AFRICAN GREEN
PIGEON

8.
ROSS'S TURACO

9.
COPPER-THROATED
SUNBIRD

10.
ORANGE-BREASTED
SUNBIRD

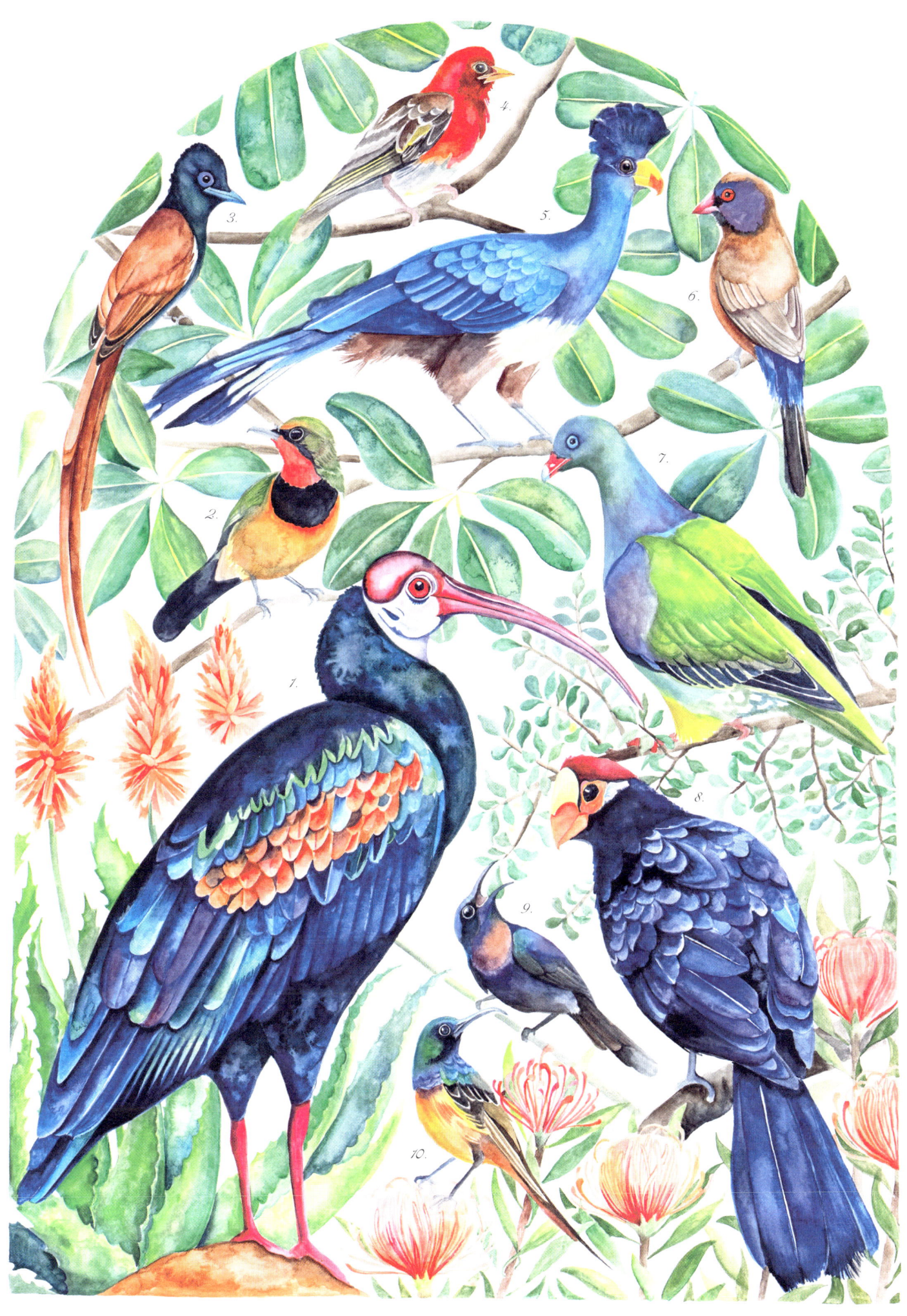

1.
SOUTHERN BALD IBIS

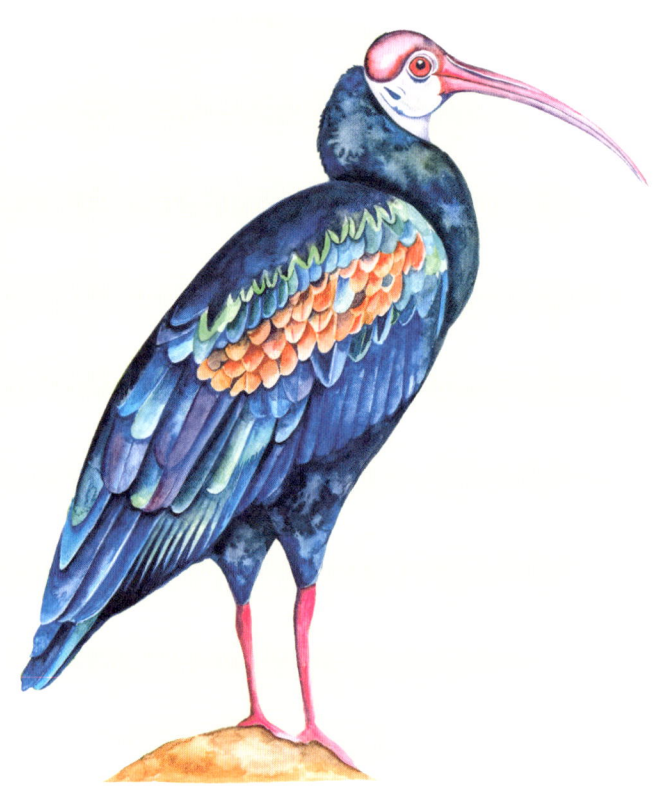

Geronticus calvus
Order Pelecaniformes

This large bird's Latin name is from the
Greek for "bald old man," owing to the bird's
featherless, bald head and wrinkled face.

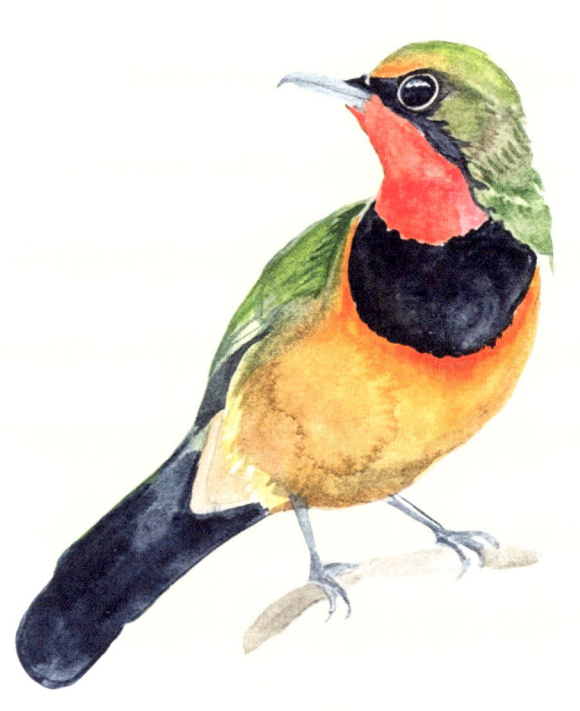

2.
GORGEOUS
BUSHSHRIKE

Telophorus viridis
Order Passeriformes

This bird is seldom seen and is known almost
exclusively by its call, which is so distinctive that the
Afrikaans word for the bird, *konkoit*, is onomatopoeic.

3.
AFRICAN PARADISE FLYCATCHER

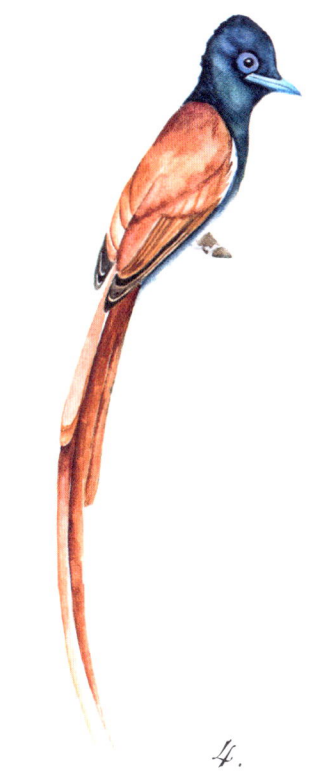

Terpsiphone viridis
Order Passeriformes

The male and female of this species share the work of building nests and incubating the eggs.

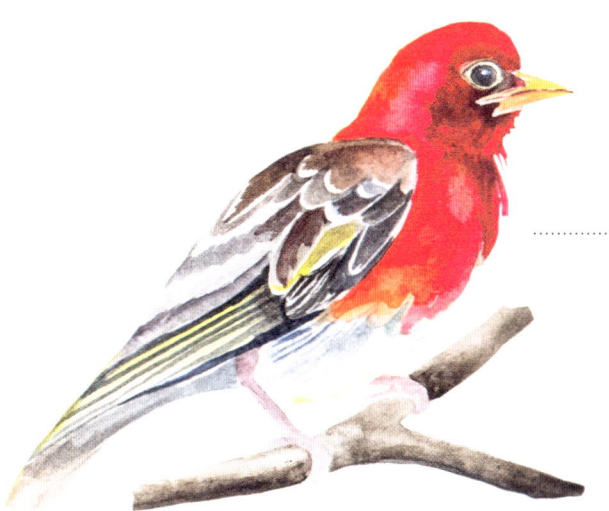

4.
RED-HEADED WEAVER

Anaplectes rubriceps
Order Passeriformes

Male weavers build suspended woven nests, often hanging on power lines or branches that reach over water. This makes it difficult for predators such as snakes to reach the nestlings.

5.
GREAT BLUE TURACO

Corythaeola cristata
Order Musophagiformes

Although they can fly, Great Blue Turacos are not great fliers. They typically hop or glide over short distances between branches.

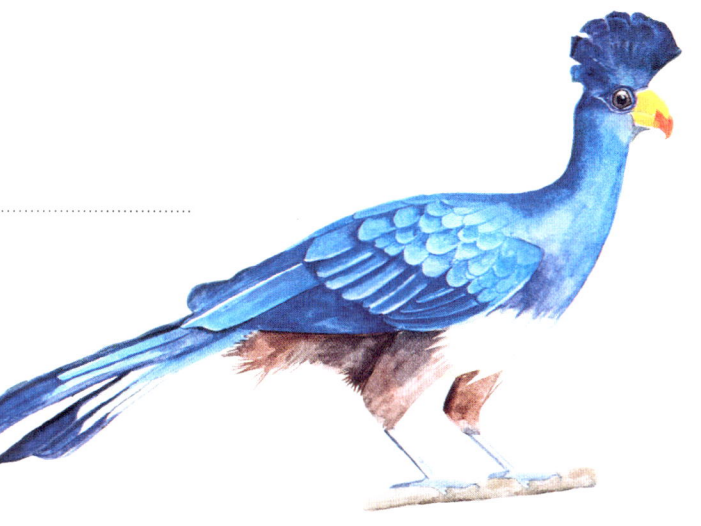

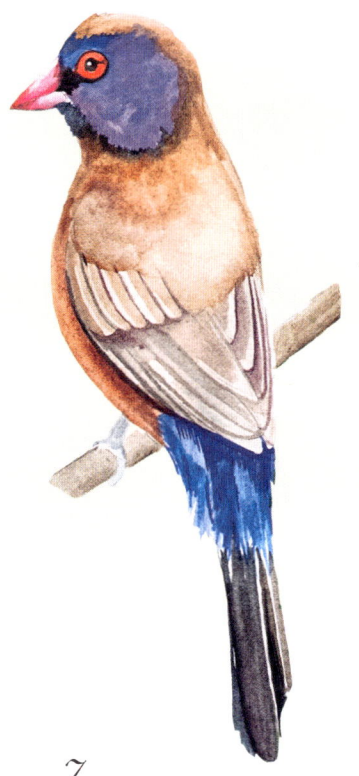

VIOLET-EARED WAXBILL

Granatina granatina
Order Passeriformes

After the Shaft-tailed Whydah lays its eggs in the nests of the Violet-eared Waxbill, the waxbill incubates the other bird's eggs and raises its chicks as though they were its own.

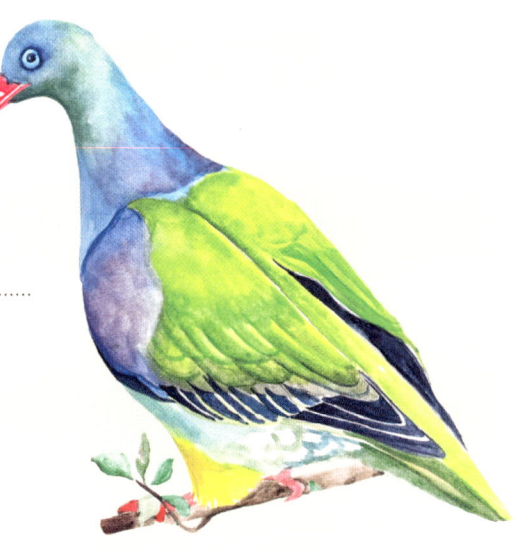

7.

AFRICAN GREEN PIGEON

Treron calvus
Order Columbiformes

These climbing birds sometimes hang upside down in the trees as they seek out hard-to-reach fruit.

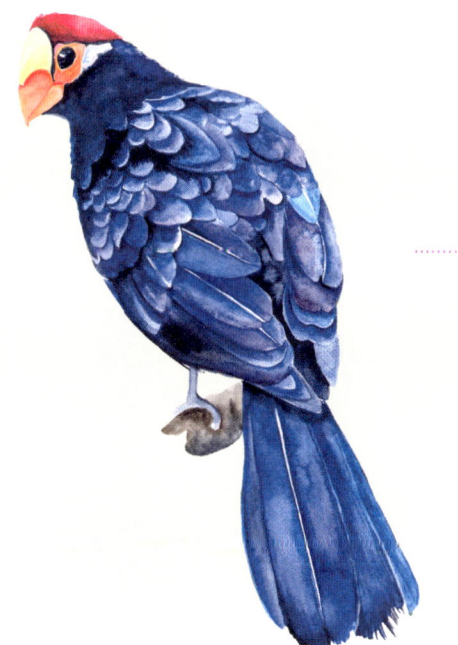

8.

ROSS'S TURACO

Musophaga rossae
Order Musophagiformes

Often considered pests in their native habitat, these birds are known to destroy gardens while searching for food.

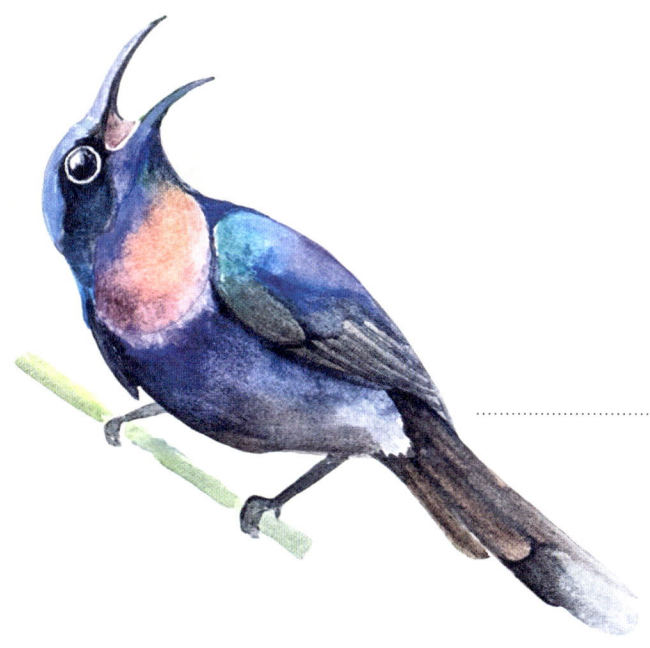

9.
COPPER-THROATED
SUNBIRD

....

Leptocoma calcostetha
Order Passeriformes

A distinctive curved bill helps this iridescent
sunbird reach flower nectar, its preferred food.

10.
ORANGE-BREASTED
SUNBIRD

....

Anthobaphes violacea
Order Passeriformes

Endemic to the mountainous fybos region of
southwestern South Aftrica, this sunbird migrates
altitudinally throughout the year, following the
blooming flowers that provide its favorite nectar.

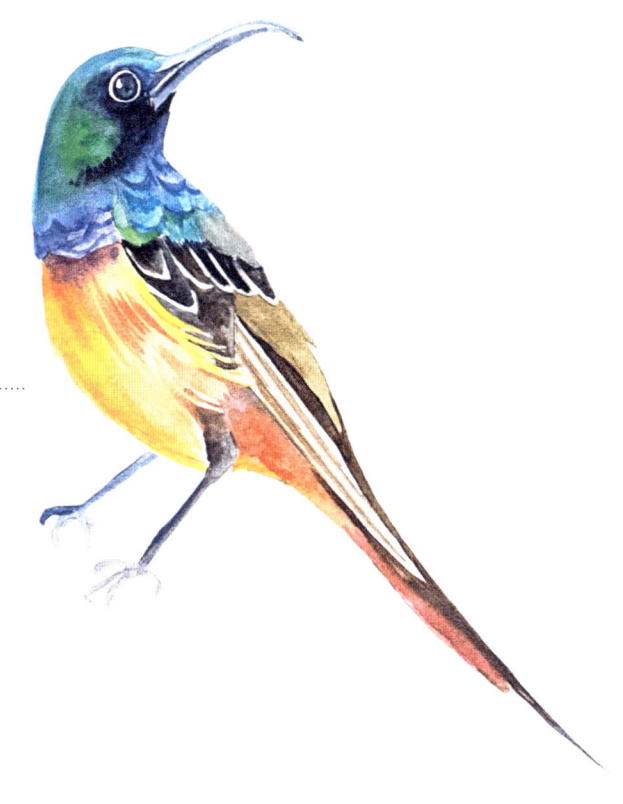

BIRDS
of
SOUTH AFRICA

✳

Whether you're traveling through South Africa's rolling hills of farmland dotted with livestock and eucalyptus or looking out at the ocean from the rugged coastline surrounded by protea blossoms, you are likely to encounter many interesting birds. Though many of the most widely known South African birds are quite large, such as the Blue Crane (the country's national bird) and the ostrich (the largest bird on Earth), we'll focus here on smaller, yet no less impressive, birds.

Sunbirds are small with characteristic curved bills, and they look similar to distantly related hummingbirds. Like hummingbirds, sunbirds have brilliantly colored feathers, and they feed on the nectar of luscious blooms. Unlike hummingbirds, which beat their tiny wings to stay aloft as they sip on nectar, sunbirds perch atop the blooms as they feed. They build small, pendulous nests of matted plant fiber that are suspended near the ground. Numerous sunbird species are found in South Africa, including the Scarlet-chested Sunbird (1) and the Malachite Sunbird (2), which can often be spotted perched atop succulent protea flowers.

Another colorful family of small birds common in South Africa are the bee-eaters, including the European Bee-eater (4), the White-throated Bee-eater (5), and the Southern Carmine Bee-eater (6). As their names suggest, these birds feed primarily on bees and can pose a significant threat to beekeepers and bee colonies. Ancient Greek philosopher Aristotle even mentioned these birds in his writings, warning beekeepers to kill them to protect their hives.

A group of birds with particularly far-reaching cultural impact are the African and Eurasian hoopoes. The African Hoopoe (10) is relatively common in South Africa, but some hoopoe species are found throughout Europe, Asia, and parts of the Middle East. They were revered in ancient Egypt, where they appear on art, tombs, and ceramics. Considered sacred and virtuous in Persian culture, the hoopoe appears as the king of all birds in *The Conference of the Birds*, a twelfth-century mystical poem considered one of the most significant of all works of Persian literature. In South Africa, you might catch an African Hoopoe's spectacular black-and-white wings as it takes flight off a fencepost or a low-hanging branch.

1.
SCARLET-CHESTED
SUNBIRD

2.
MALACHITE SUNBIRD

3.
CAPE SUGARBIRD

4.
EUROPEAN
BEE-EATER

5.
WHITE-THROATED
BEE-EATER

6.
SOUTHERN CARMINE
BEE-EATER

7.
CRIMSON-BREASTED
GONOLEK

8.
CRESTED BARBET

9.
LILAC-BREASTED
ROLLER

10.
AFRICAN
HOOPOE

11.
CAPE WHITE-EYE

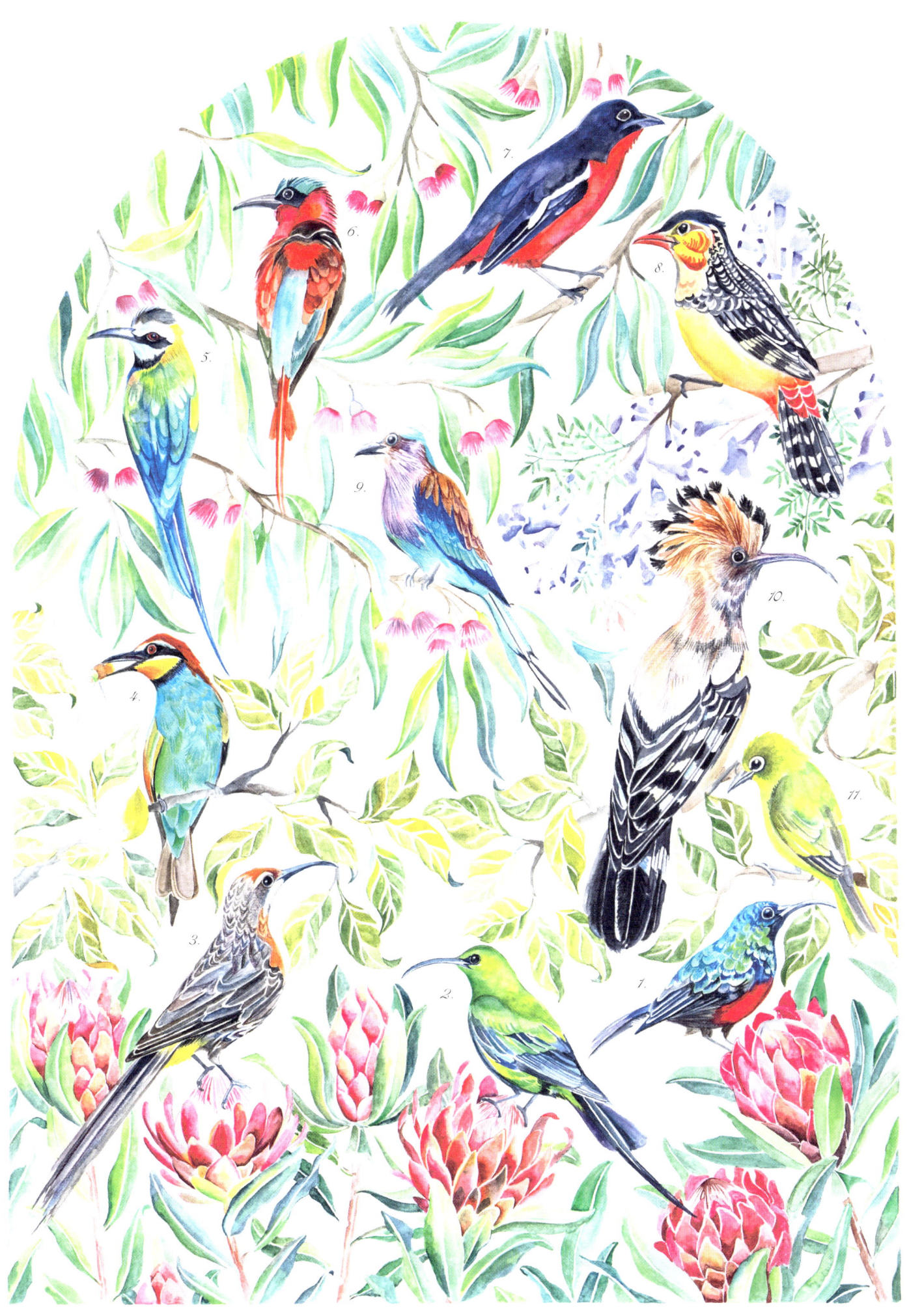

1.
SCARLET-CHESTED
SUNBIRD

Chalcomitra senegalensis
Order Passeriformes

This brightly colored bird is easily
confused with the Hunter's Sunbird,
which looks remarkably similar.

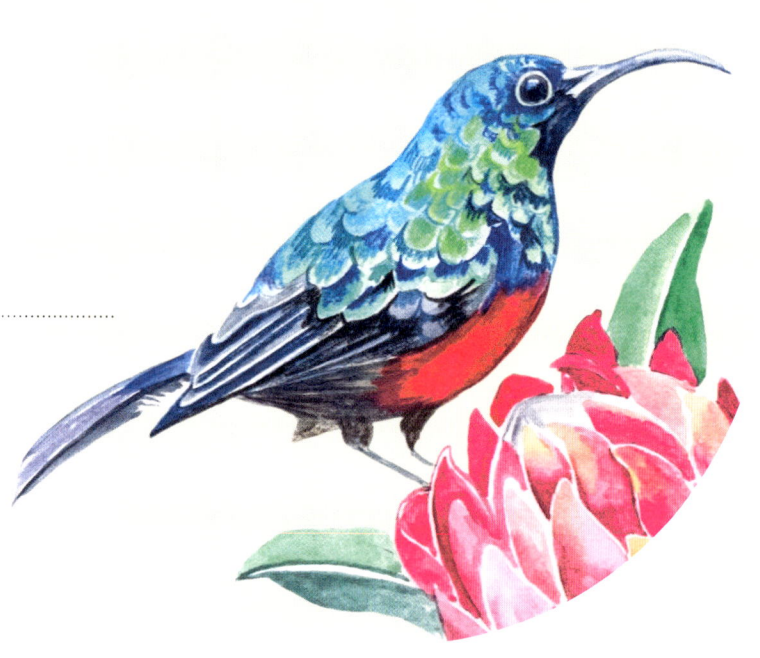

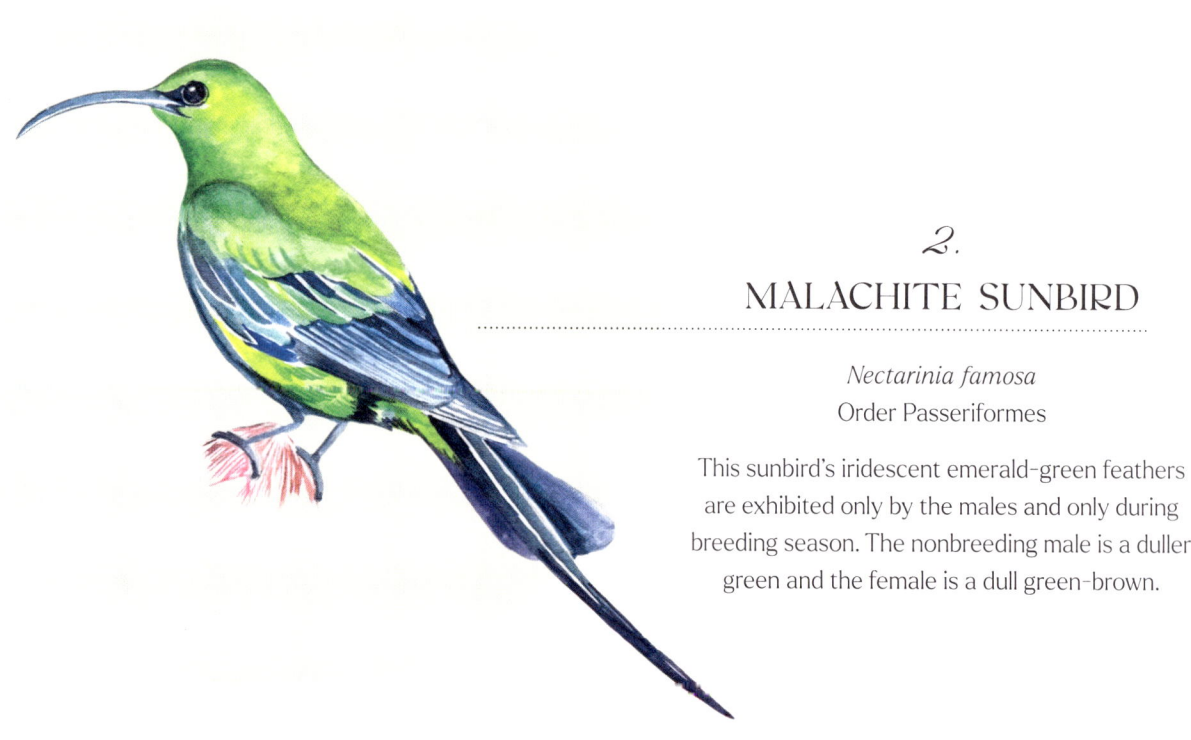

2.
MALACHITE SUNBIRD

Nectarinia famosa
Order Passeriformes

This sunbird's iridescent emerald-green feathers
are exhibited only by the males and only during
breeding season. The nonbreeding male is a duller
green and the female is a dull green-brown.

3.
CAPE SUGARBIRD

Promerops cafer
Order Passeriformes

Endemic to South Africa, Cape Sugarbirds
feed on the nectar of the protea,
the country's national flower.

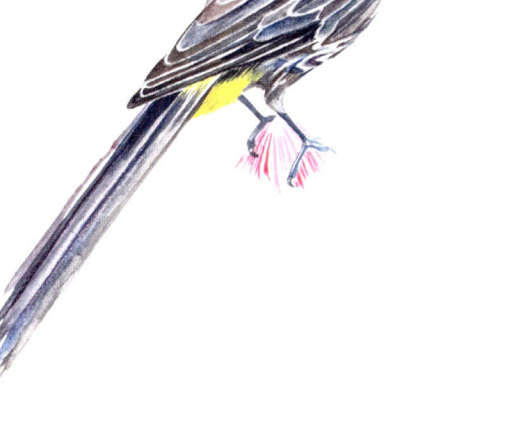

4.
EUROPEAN BEE-EATER

Merops apiaster
Order Coraciiformes

This bee-eater has an incredibly large
range that extends across Africa, Europe,
and parts of the Middle East.

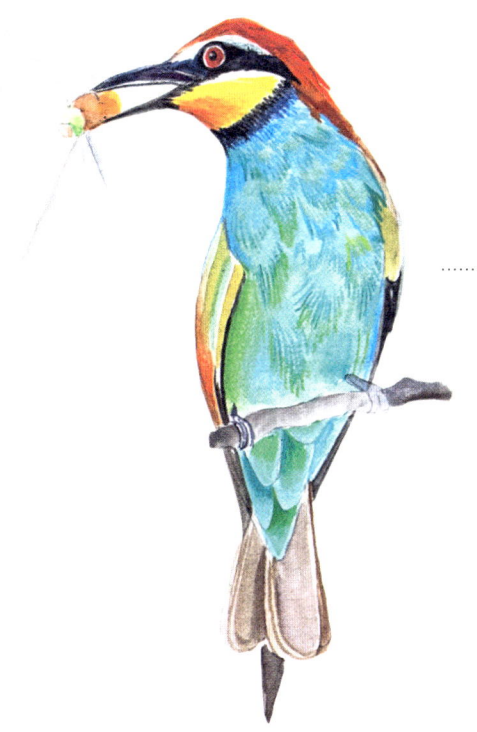

5.
WHITE-THROATED
BEE-EATER

Merops albicollis
Order Coraciiformes

Like other bee-eaters, these birds build tunnel-like
nests in sandy riverbanks or other open sandy areas.

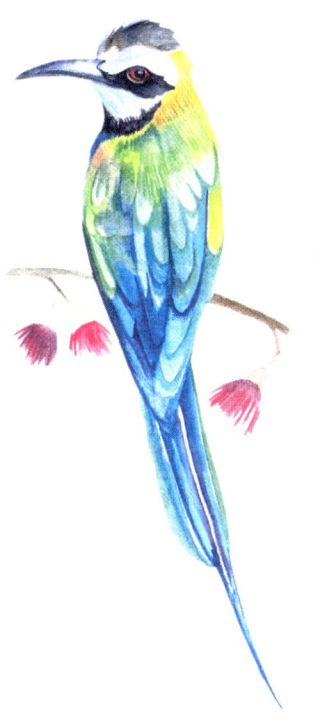

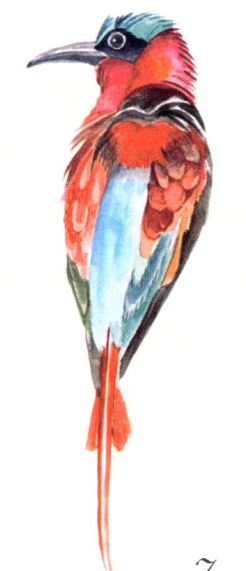

6.
SOUTHERN CARMINE BEE-EATER

Merops nubicoides
Order Coraciiformes

These birds snatch insects from the air after launching from a perch. They are known to perch on branches—or even the backs of other animals—before flying off to grab prey.

7.
CRIMSON-BREASTED GONOLEK

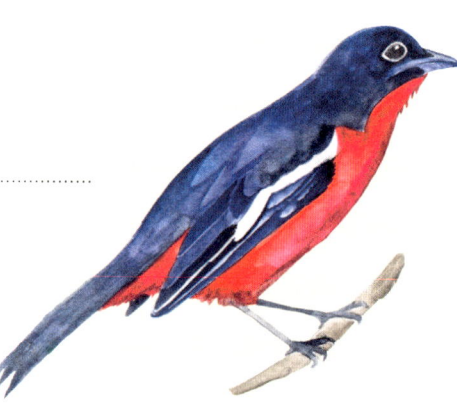

Laniarius atrococcineus
Order Passeriformes

Because this bird's striking red feathers reminded German colonialists of their native flag, it was nicknamed *reichsvogel*, or "empire bird."

8.
CRESTED BARBET

Trachyphonus vaillantii
Order Piciformes

These highly territorial birds will attack other birds and small animals if they view them as competitors or threats.

9.
LILAC-BREASTED ROLLER

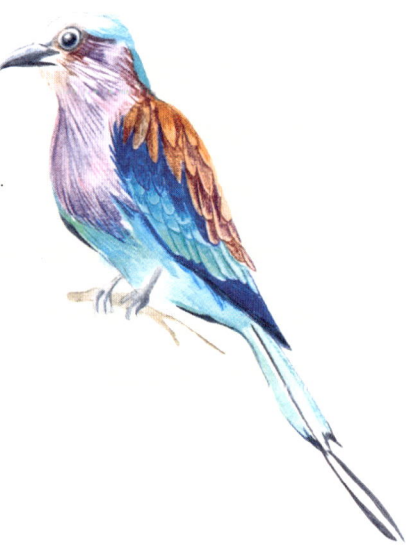

Coracias caudatus
Order Coraciiformes

The roller is named for its courtship display, which involves diving downward in flight while rolling its body.

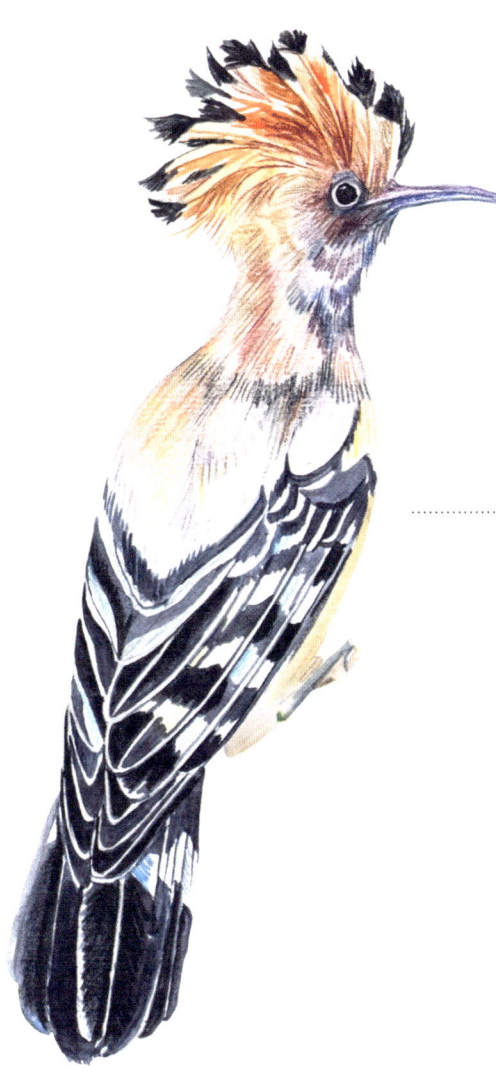

10.
AFRICAN HOOPOE

Upupa africana
Order Bucerotiformes

Once considered the same species as the similar-looking Eurasian Hoopoe (*U. epops*), this bird was later categorized as a separate species.

11.
CAPE WHITE-EYE

Zosterops virens
Order Passeriformes

These little warblers are a birdwatching favorite because of their beautiful and distinctive call.

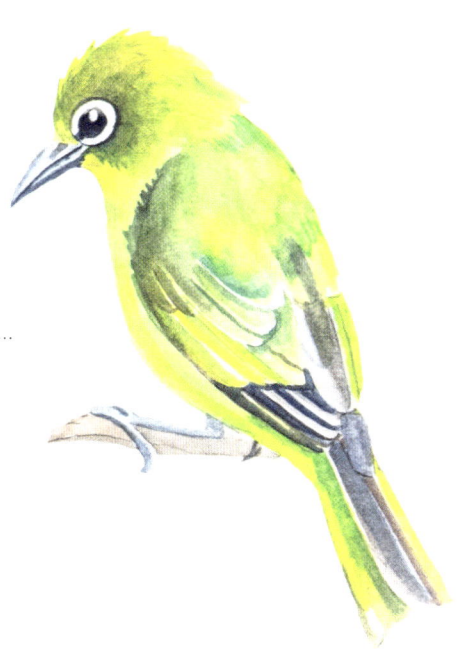

BIRDS
of the
EAST AFRICAN
SAVANNA

A chance to see a Gray Crowned-Crane (3)—the large, graceful wetland dweller—is reason enough to go bird-watching on the African savanna. Look for its unmistakable mohawk-like crown of yellow feathers and long, purple-grey body. It exhibits peculiar feeding habits and sometimes stomps its feet to provoke insects to take flight before snatching them from the air. Considered a symbol of hope and grace in several African countries, this crane is an endangered species, but huge efforts are being made to protect these spectacular birds.

Another bird with unusual habits is the Red-billed Oxpecker (5), commonly seen perched on the backs of livestock and other large ungulates, including impalas, kudus, and sometimes even rhinoceroses. This bird has a symbiotic relationship with these mammals and picks off insects from their hosts' skin—although they can sometimes pick too vigorously, resulting in small wounds and sores. The Swahili name for this bird, *askari wa kifaru*, translates to "the rhino's guard": the bird becomes agitated when it senses that a predator is near, alerting its animal host to approaching danger.

The Bateleur (6) is a member of a group often referred to as the "serpent eagles." They are known to eat snakes, along with a variety of other prey, and although they generally do their own hunting and killing, they occasionally dine on carrion. A large, impressive raptor with broad, dark wings and a bald red face, this eagle has inspired and entranced various African cultures. The flag and currency of the country of Zimbabwe shows the symbol of a bird of prey largely thought to represent the Bateleur, and the bird appears in several myths from South Africa.

1.
BLACK-HEADED GONOLEK

2.
ZEBRA WAXBILL

3.
GRAY CROWNED-CRANE

4.
RED-BILLED FIREFINCH

5.
RED-BILLED OXPECKER

6.
BATELEUR

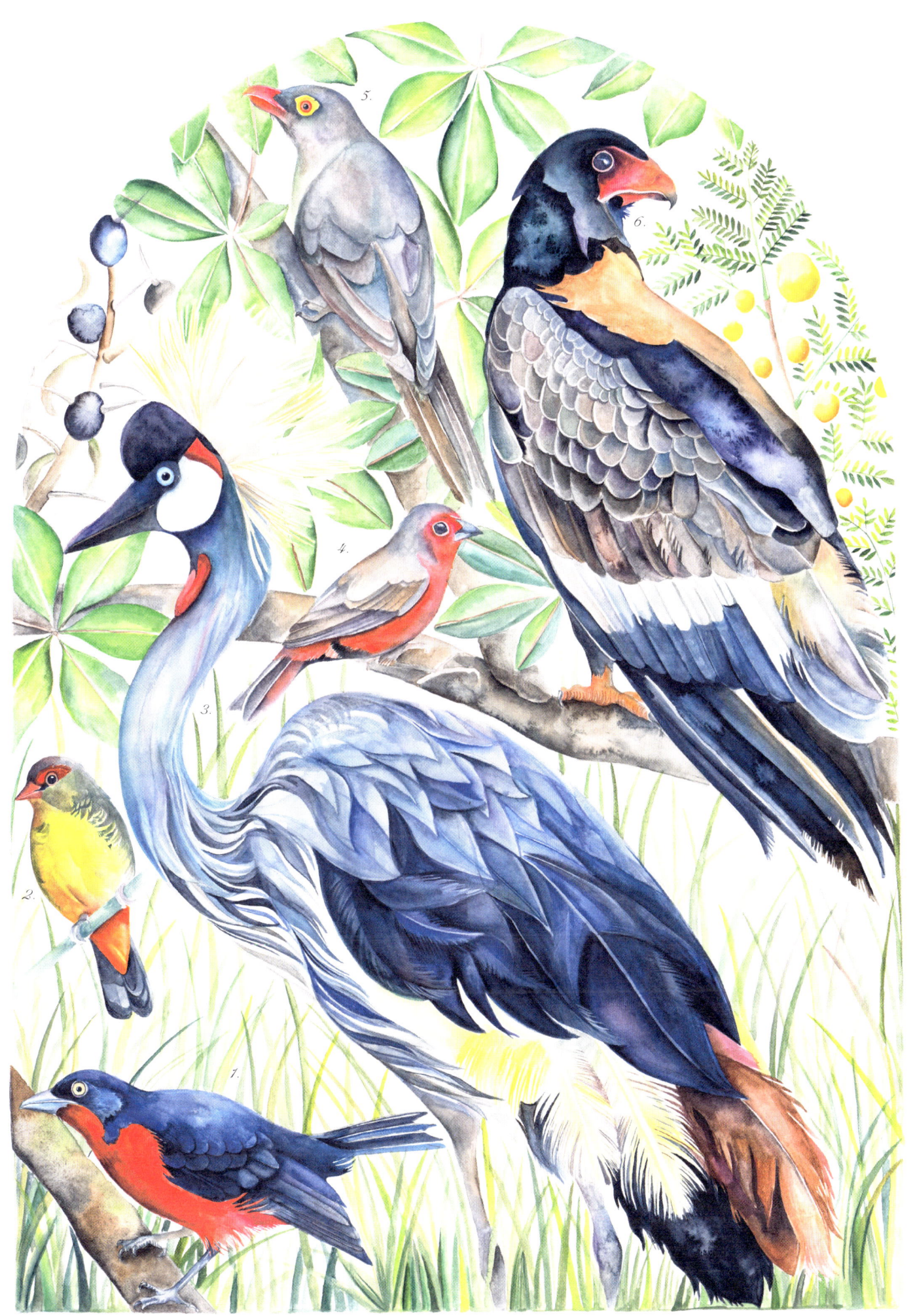

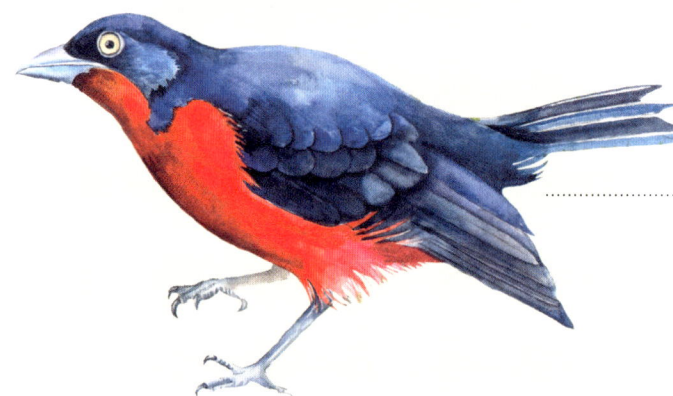

1.
BLACK-HEADED GONOLEK

Laniarius erythrogaster
Order Passeriformes

The vibrantly colored gonolek is generally a shy and skulking bird of grasslands and shrublands, but those that nest near people can become quite bold.

2.
ZEBRA WAXBILL

Amandava subflava
Order Passeriformes

Sometimes known as the Orange-breasted Waxbill, this sparrow-sized bird has an orange breast and dark striped patterning on its flanks.

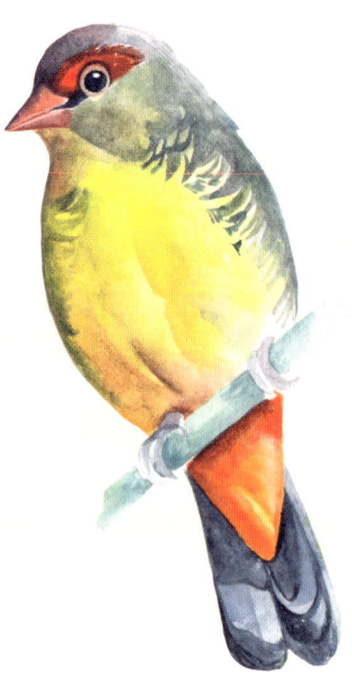

3.
GRAY CROWNED-CRANE

Balearica regulorum
Order Gruiformes

The national bird of Uganda, this large, crested bird is featured at the center of the country's flag.

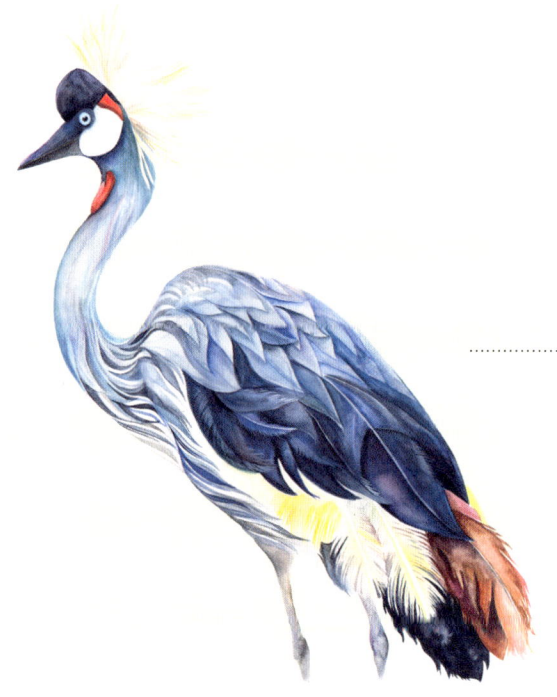

4.
RED-BILLED FIREFINCH

Lagonosticta senegala
Order Passeriformes

The Village Indigobird, a nest parasite, lays its eggs in a Red-billed Firefinch's nest, where the nestlings are reared by the host species along with their own young.

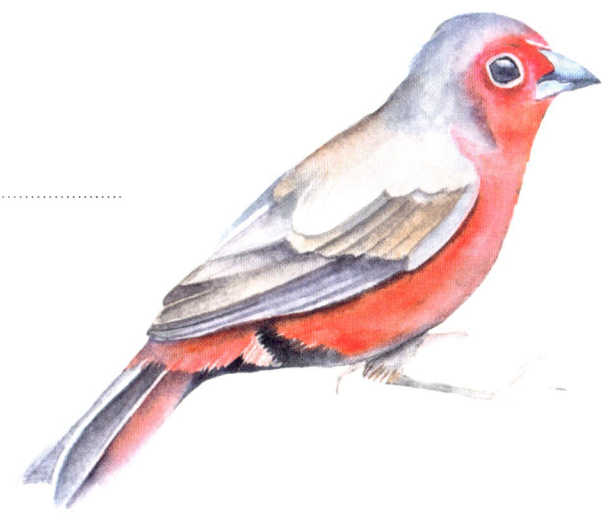

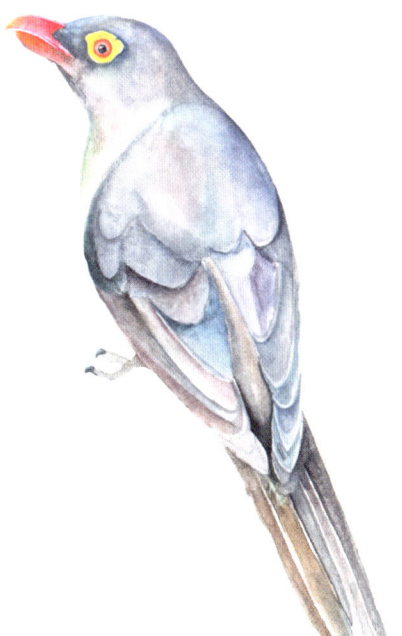

5.
RED-BILLED OXPECKER

Buphagus erythrorynchus
Order Passeriformes

As the oxpecker pecks insects from the skin of its ungulate host, it sometimes opens wounds on the animal. The bird will sometimes feed on the blood from the open wound.

6.
BATELEUR

Terathopius ecaudatus
Order Accipitriformes

When this short-tailed eagle is on the hunt, it can fly at incredible speeds. It sometimes performs impressive acrobatics as it snatches prey in midair.

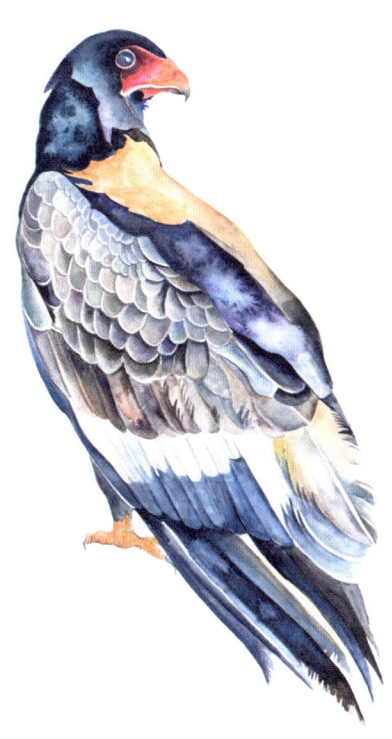

BIRDS
of the
NILE RIVER
VALLEY

Egypt is a land of natural extremes. The Nile, the longest river in Africa, cuts a path through the desert, carving a line of lush, fertile land into the parched and forbidding areas that surrounds it. Many birds make their homes along the fecund banks of the Nile among the reeds and trees and in the areas of adjacent farmland. Among the grasses, you may catch the glinting iridescent feathers of the tiny Nile Valley Sunbird (2). Among the abundant papyrus marshes, look for the striking blue body of an African Swamphen (7).

Birds have a rich cultural history in Egypt, as they held an important place in the lore and worship of ancient Egyptians. Birds are pervasive symbols in art and religious relics from antiquity, appearing on pottery, murals, furniture, and jewelry and throughout tombs and temples. Horus, an important god in the ancient Egyptian pantheon, had the head of a falcon and the body of a man; one of its eyes represented the sun and the other the moon. The Egyptian Goose (1), which frequented the rich and fruitful lands near the Nile, were considered auspicious omens of fertility and abundance and were used in the decorative motifs of tombs as symbols of departed souls. The Egyptian Vulture (5) was frequently used as a symbol of royalty and is represented in hieroglyphics and many relics and artworks. The genus name for this bird, *Neophron*, comes from a dishonorable character in Greek mythology who is transformed into a vulture by the gods as punishment for his deceptions and indiscretions.

Many images and depictions of birds from this period show them not only as decorative and religious motifs, but as birds in action: A Pied Kingfisher (3), wings folded into his body, plunges sharply into the water. An Egyptian Goose (1) wanders in dense growths of papyrus along the riverbank. Ancient Egyptians also consumed birds and were known to raise doves for their meat and to carry messages, a tradition that continued after the fall of their civilization.

In addition to its year-round avian residents, Egypt is also on the path of a major bird migration route and plays seasonal host to several species that pass between Europe and Africa. One such bird is the European Bee-eater (6), a richly colored species that nests near (or often within) riverbanks, where it hunts for insect prey.

1.
EGYPTIAN GOOSE

2.
NILE VALLEY SUNBIRD

3.
PIED KINGFISHER

4.
LAUGHING DOVE

5.
EGYPTIAN VULTURE

6.
EUROPEAN BEE-EATER

7.
AFRICAN SWAMPHEN

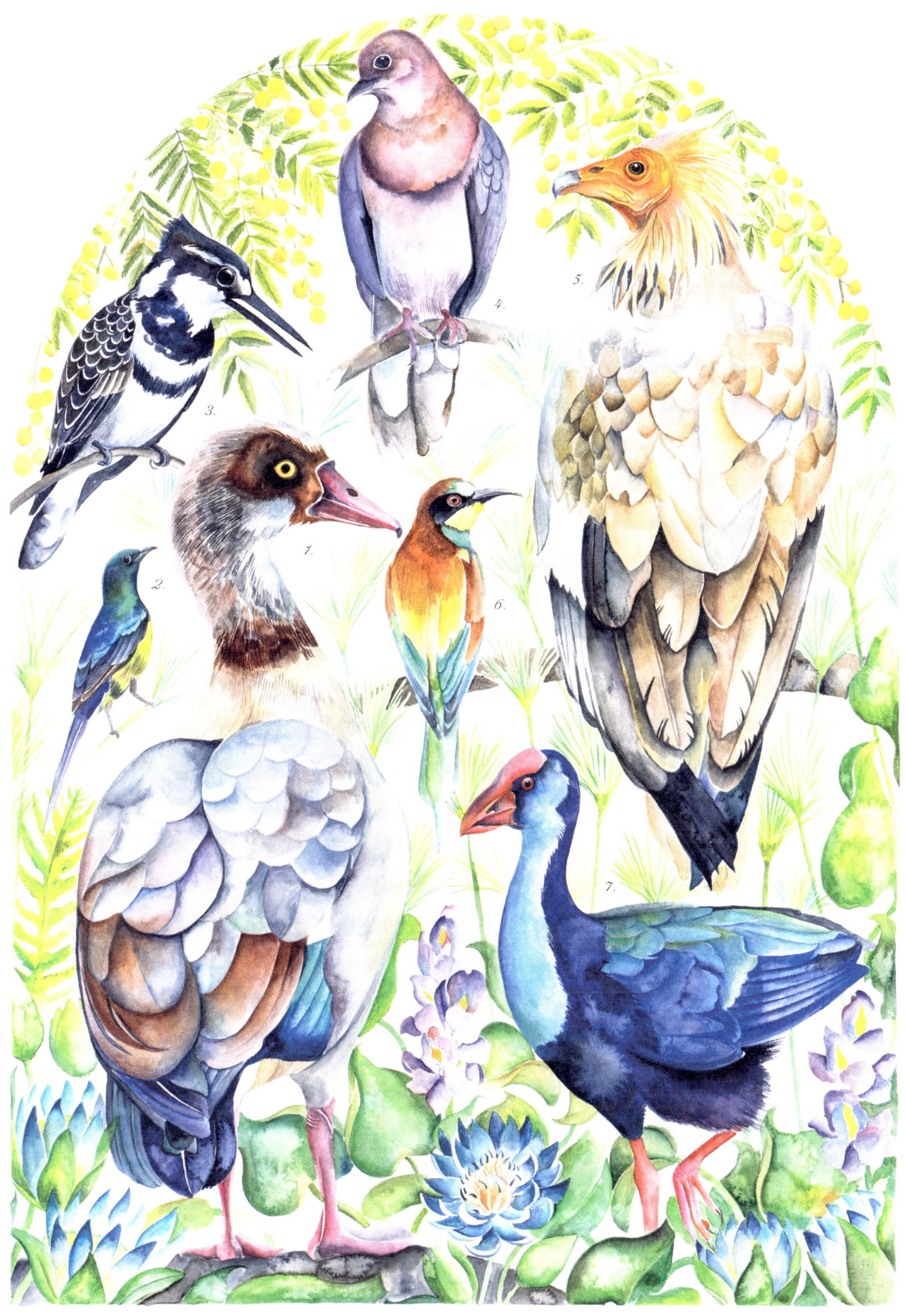

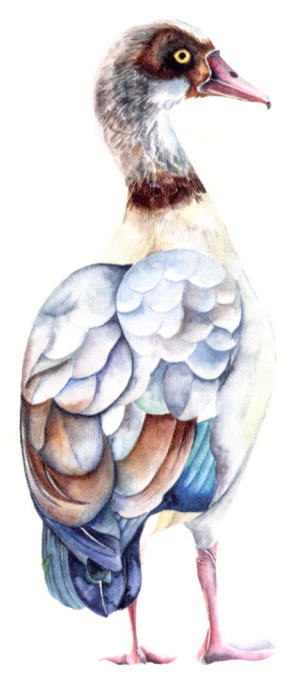

1.
EGYPTIAN GOOSE

Alopochen aegyptiaca
Order Anseriformes

Despite its common name, this waterbird is more closely related to a duck than a goose.

2.
NILE VALLEY SUNBIRD

Hedydipna metallica
Order Passeriformes

During much of the year, male and female sunbirds are relatively similar looking, but during breeding season, the male displays metallic, colorful feathers and a long, streaming tail.

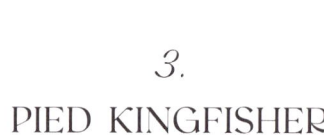

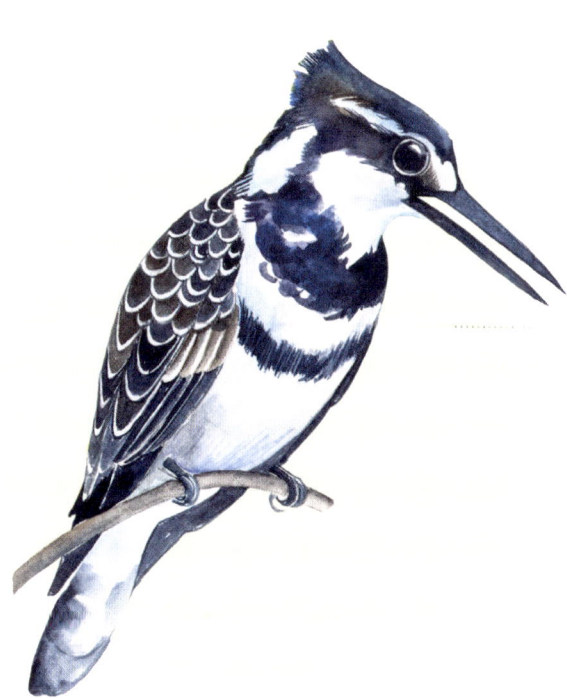

3.
PIED KINGFISHER

Ceryle rudis
Order Coraciiformes

These brilliant birds hover over the water in search of prey before diving in headfirst, with their bills pointed spear-like into the water.

4.
LAUGHING DOVE

Spilopelia senegalensis
Order Columbiformes

Usually seen in pairs, these faithful birds mate for life.

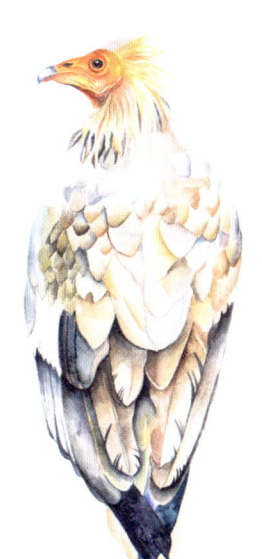

5.
EGYPTIAN VULTURE

Neophron percnopterus
Order Accipitriformes

Egyptian Vultures have been observed using tools: they use rocks to crack open large eggs and sticks to gather and roll nesting material.

6.
EUROPEAN BEE-EATER

Merops apiaster
Order Coraciiformes

This species feeds mainly on insects caught during flight. After it catches a bee, it bashes the insect against a hard surface to remove the stinger before eating it.

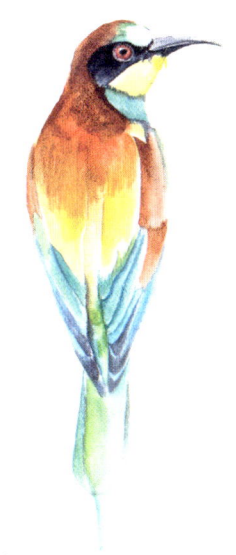

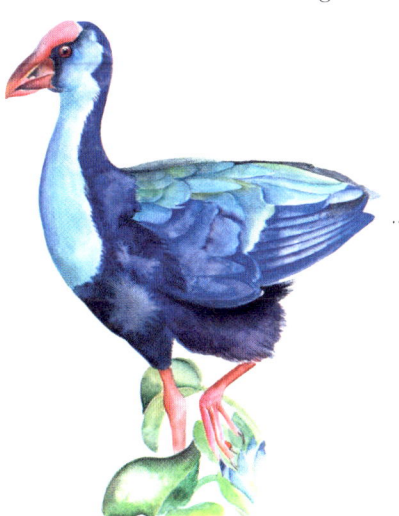

7.
AFRICAN SWAMPHEN

Porphyrio madagascariensis
Order Gruiformes

This bird was once considered a subspecies of the closely related and similar-looking Purple Gallinule, but they have since been classified as two separate species.

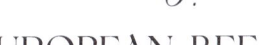

BIRDS of WEST AFRICA

Although the coast of western Africa is largely tropical, many large, open, arid regions are located farther inland. Among the region's well-known lions, elephants, and wildebeest populations are more than 1000 species of birds. For this reason, the area has become a destination for birders from around the world.

One of the most astonishing birds that's relatively common across the dry lands of the savanna is the Vulturine Guineafowl (1). This large bird has a brilliant sapphire-blue head and breast, which is adorned with spiky black-and-white feathers, like an elaborate and beautiful neckpiece. These fast runners rarely fly and can cover the ground quickly on foot. Various African guineafowl are thought to have been brought to Europe and domesticated by the ancient Greeks and Romans. These domesticated birds disappeared from the continent with the fall of the Roman Empire and weren't seen again in Europe until the fifteenth century, when Portuguese explorers brought several species of guineafowl back to Europe with them.

Birders may come across a Northern Red-billed Hornbill (3), which serves as a symbol of faith and positivity in various cultures of the region. Female hornbills have a unique strategy for rearing and safeguarding their young. After she chooses a tree cavity for nesting, she seals the entrance from the inside using her own droppings and other collected natural materials to protect herself and her babies from intruders. When the male brings food for his family, he transfers it through a tiny opening created by the female for this purpose.

The Hadada Ibis (5) is a beautiful but common bird that is considered a pest by locals for its habit of perching on rooftops and making extremely loud *ha-da-da* calls early in the morning (hence the common name). In Lesotho, some people believe that the call of this bird foretells coming rains, and various tribes and cultures throughout southwestern Africa share folk tales about this well-known bird.

1.
VULTURINE GUINEAFOWL

2.
YELLOW-NECKED SPURFOWL

3.
NORTHERN RED-BILLED HORNBILL

4.
WHITE-FRONTED BEE-EATER

5.
HADADA IBIS

6.
ABYSSINIAN ROLLER

7.
BRONZE SUNBIRD

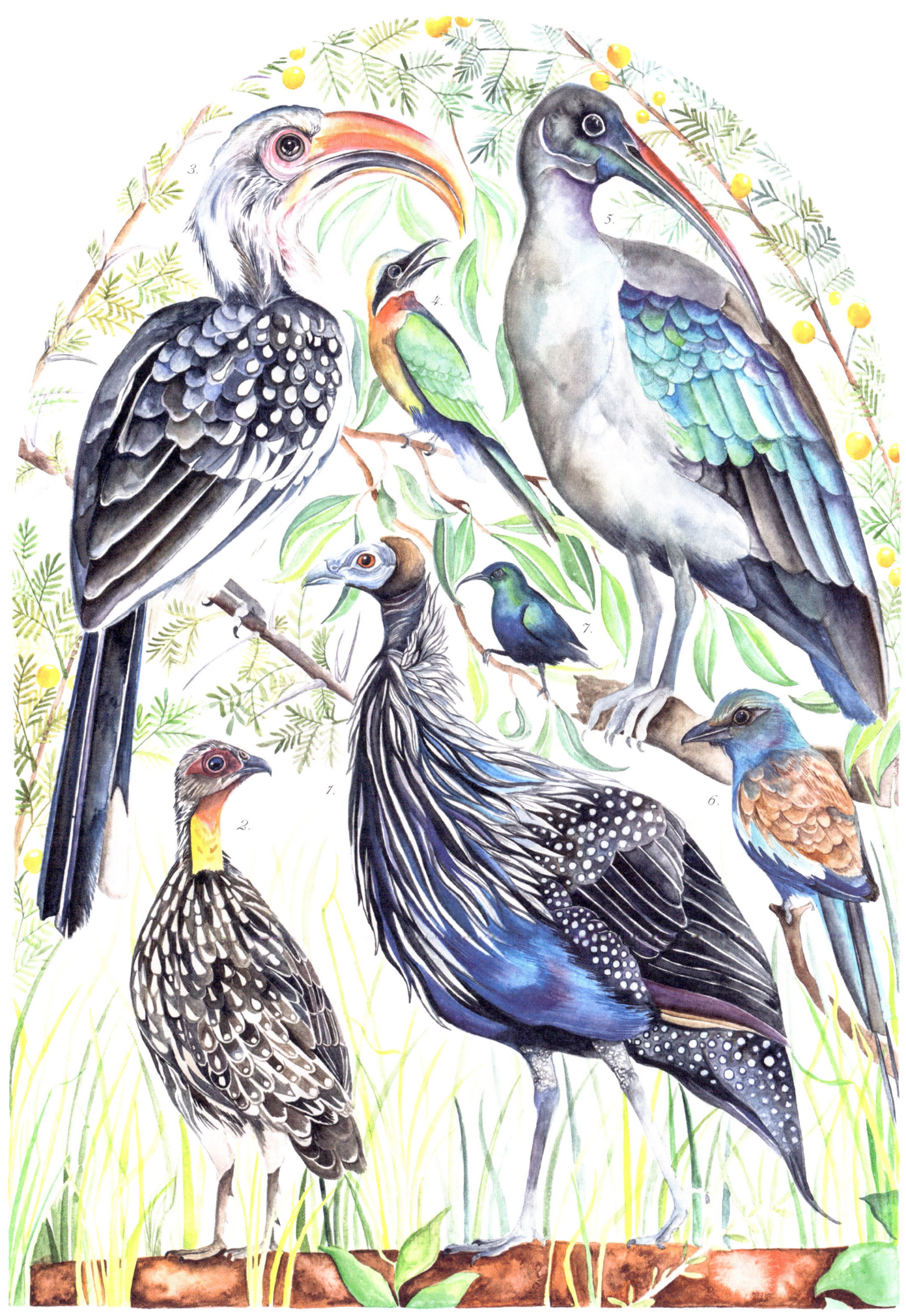

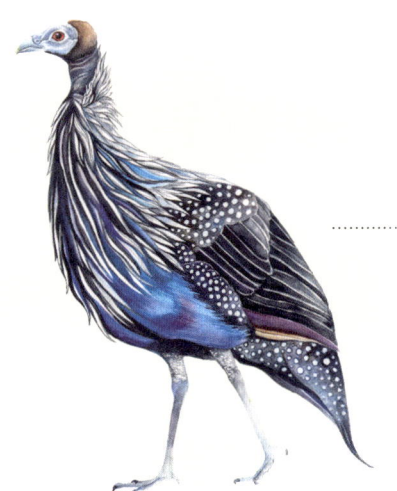

1.
VULTURINE GUINEAFOWL

Acryllium vulturinum
Order Galliformes

Outside the breeding season, these birds often travel
in flocks of typically about twenty-five birds.

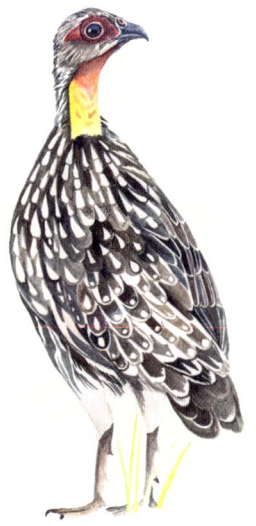

2.
YELLOW-NECKED SPURFOWL

Pternistis leucoscepus
Order Galliformes

Found in savanna habitats and in adjacent agricultural
fields, this highly adaptable bird will tolerate some
agricultural development within its territory.

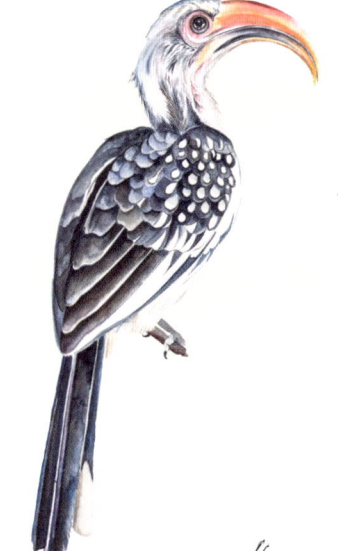

3.
NORTHERN RED-BILLED HORNBILL

Tockus erythrorhynchus
Order Bucerotiformes

This omnivore often visits piles of droppings left by large
mammals to pick out and eat beetles and fly larvae.

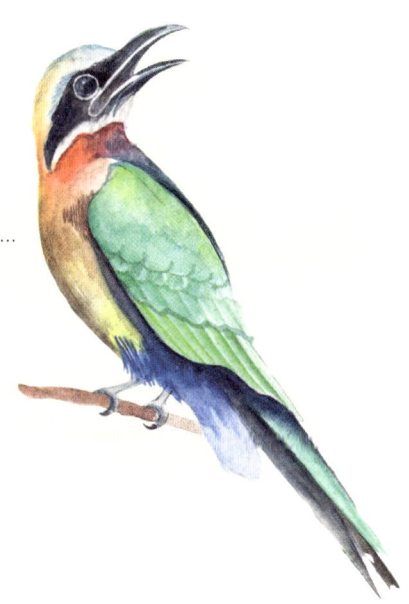

4.
WHITE-FRONTED BEE-EATER

Merops bullockoides
Order Coraciiformes

These birds have a complex social dynamic within their species and
build nesting colonies of up to 450 birds with a hierarchy of clans.

5.
HADADA IBIS

Bostrychia hagedash
Order Pelecaniformes

These birds often forage in agricultural fields, using their scythe-like bills to pick insects from the ground. Farmers often welcome their foraging, as the birds often eat insects that would otherwise damage crops, and the small holes they leave in the ground help aerate the soil.

7.
BRONZE SUNBIRD

Nectarinia kilimensis
Order Passeriformes

This bird has benefited from the proliferation of cultivated protea flowers in Africa. These large, striking flowers are poisonous to humans but provide vital nectar to the Bronze Sunbird.

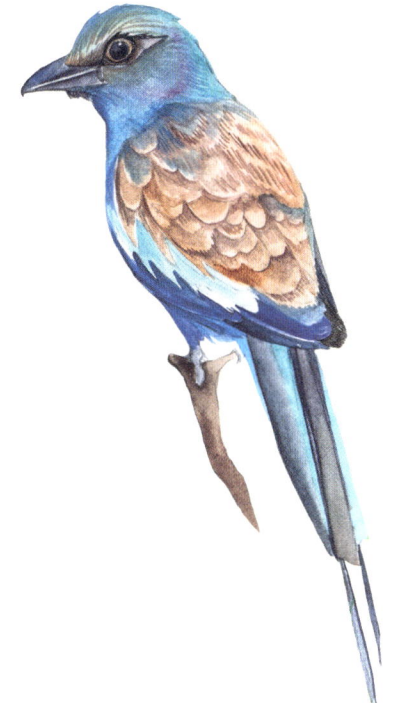

6.
ABYSSINIAN ROLLER

Coracias abyssinicus
Order Coraciiformes

These fearless birds feed on mostly invertebrates. They have been observed diving into forest fires to hunt species provoked out of hiding by the smoke and flames.

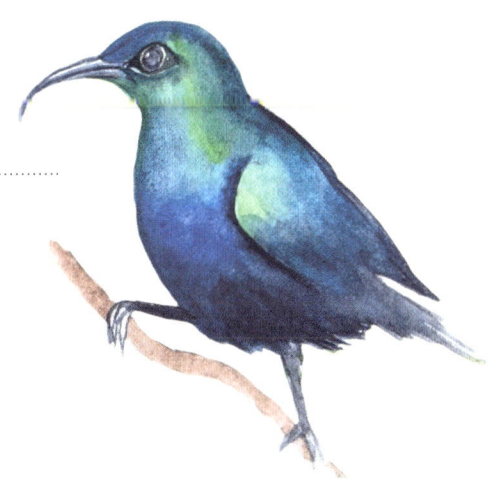

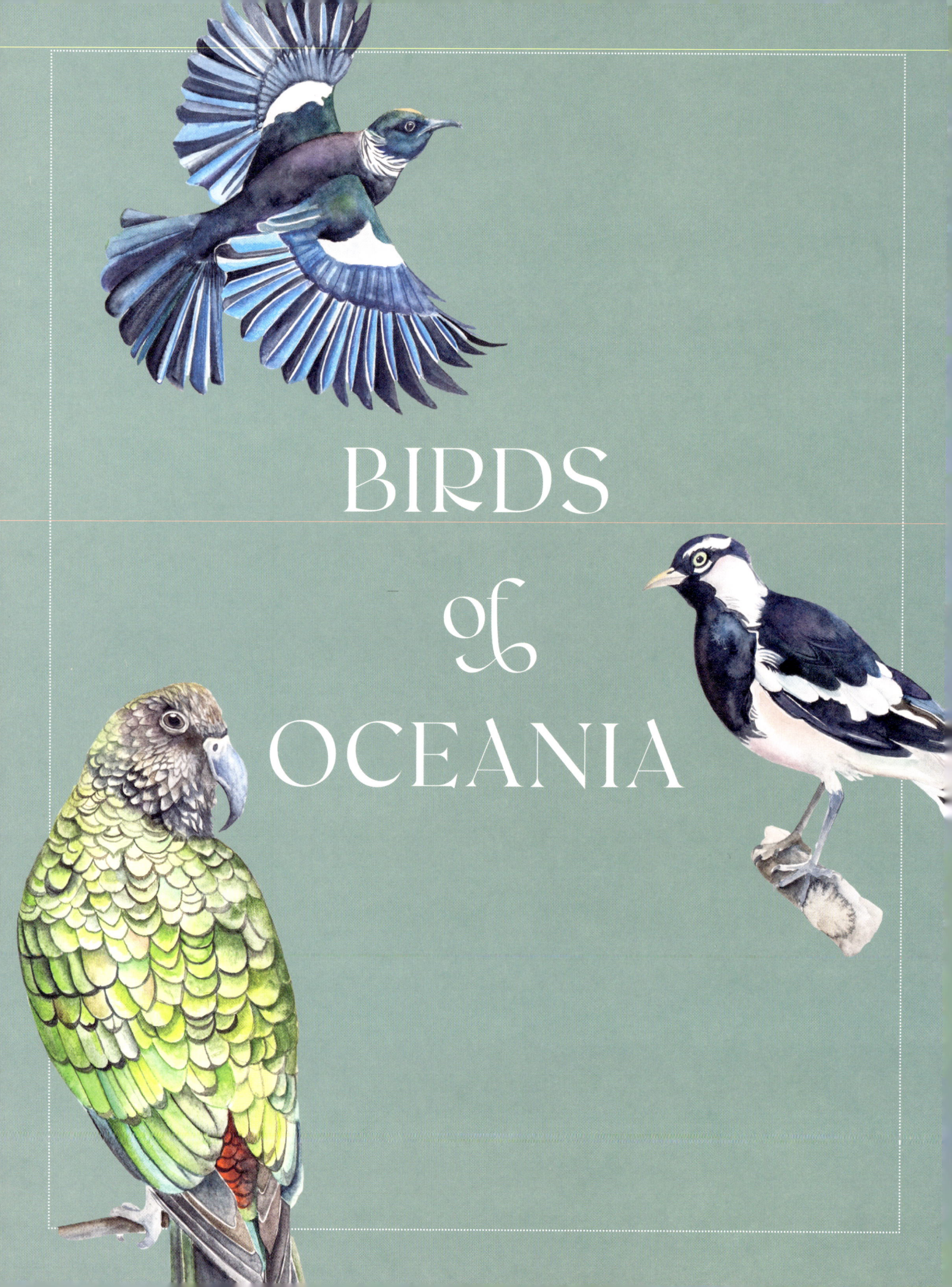

BIRDS
of
OCEANIA

This geographical region includes many island nations, including Australia and New Zealand, and is home to many beautiful and varied landscapes and about 2000 species of birds. Because of their relative isolation, many of the islands have an exceptionally high number of endemic bird species. Here we will explore some of the notable endemics from New Zealand and Hawaii as well as some of the tropical avian jewels of Papua New Guinea.

Endemic New Zealand birds include the Kea, the only living species of mountain parrot, and the New Zealand Pigeon (*kererū* in the Māori language), which is known to get drunk after feasting on fermented berries. You may be familiar with some of the parrots, parakeets, and lorikeets native to Australia, as today many are commonly kept as pets throughout the world.

BIRDS
of
NEW ZEALAND

Home to lush, fern-covered valleys, verdant mossy forests, thundering waterfalls, glistening white-sand beaches, and soaring ice-capped mountain peaks, New Zealand has one of the most unusual ecosystems in the world. Because of its geographic isolation in the South Pacific Ocean, it is host to numerous endemic bird species that could survive nowhere else. Prior to human settlement, no mammals lived on these islands, which enabled a vast number of unusual bird species to thrive without natural predators. Tragically, after the introduction of several invasive mammalian predators and overdevelopment by humans, much of the bird life in New Zealand is now critically endangered. Massive conservation efforts are underway and have had some success in slowing population declines among native birds.

Probably the most common bird illustrated here, and certainly the one you are most likely to encounter, is the New Zealand Fantail (4), a small endemic species that is a common sight along forest trails and in urban parks. It is easily identified by the graceful arch of its fanned tail, for which it is named.

In the backcountry, hikers may catch the iridescent gleam of the large New Zealand Pigeon, or *kererū* (6), whose glinting green feathers can be seen flickering through the leaves as it swoops between trees. These colorful birds are famously clumsy and have been known to fall out of trees after getting drunk by over-imbibing on fermented berries.

Backcountry hikers on New Zealand's South Island may also encounter the Kea (8), the only species of alpine parrot in the world. In Māori legend, these birds, named for the call they make in flight, are considered guardians of the mountains. The cheeky birds, sometimes called "clowns of the mountains," are highly social and are known as curious pests to hikers because of their tendency to investigate unguarded backpacks and clothing. They may even fly off with goods stolen from unsuspecting hikers. Watch out!

The Māori people hold birds in extremely high regard and watch their movements and behaviors carefully, believing that the activities of certain birds predict weather patterns and seasonal changes. For example, in spring, the arrival of the Shining Bronze-Cuckoo (9) indicates that the time has come to plant crops.

1.	3.	5.	7.	9.
NORTH ISLAND SADDLEBACK	TŪĪ	STITCHBIRD (HIHI)	SACRED KINGFISHER	SHINING BRONZE-CUCKOO
2.	4.	6.	8.	10.
NEW ZEALAND BELLBIRD	NEW ZEALAND FANTAIL	NEW ZEALAND PIGEON (KERERŪ)	KEA	TŪĪ

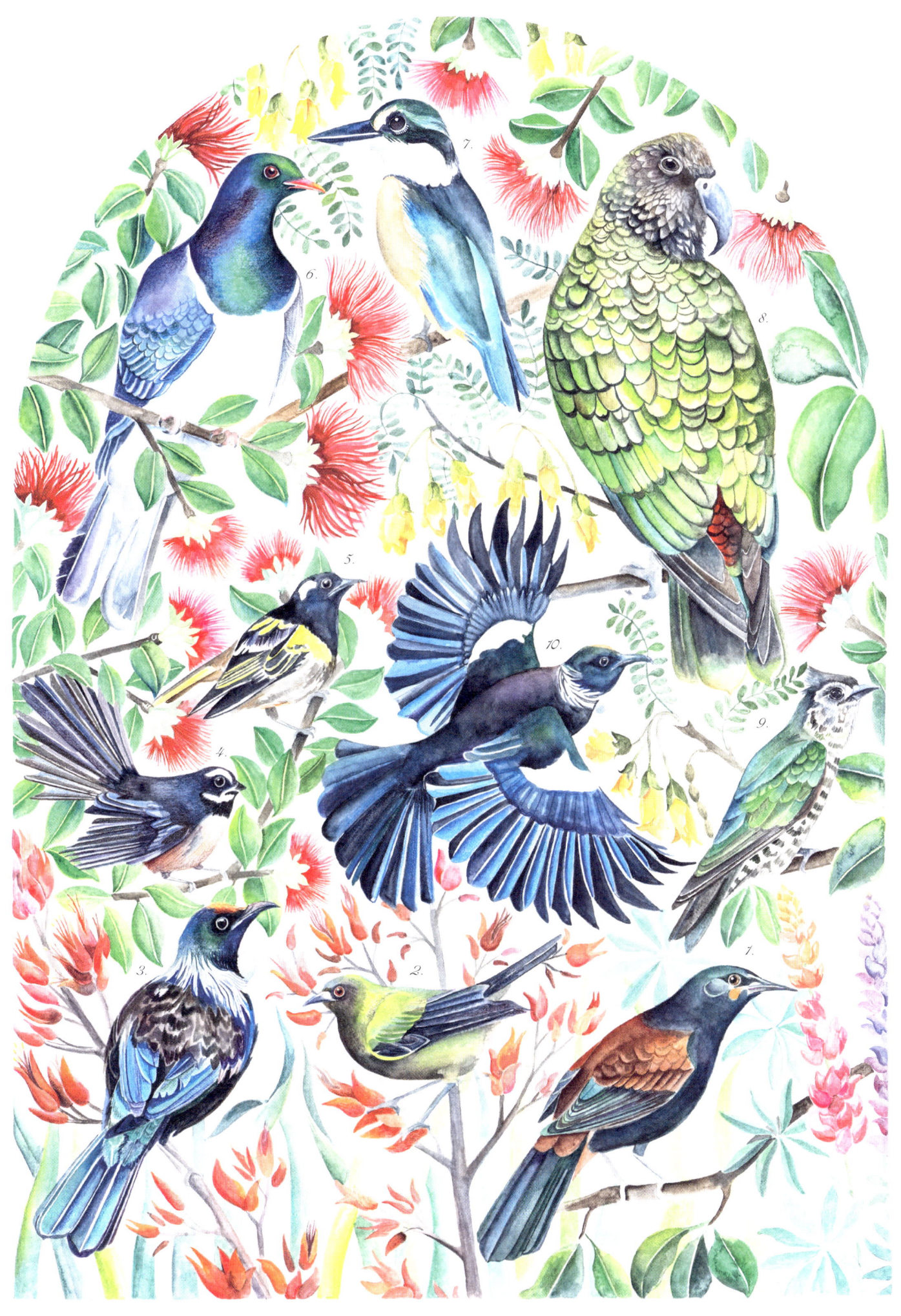

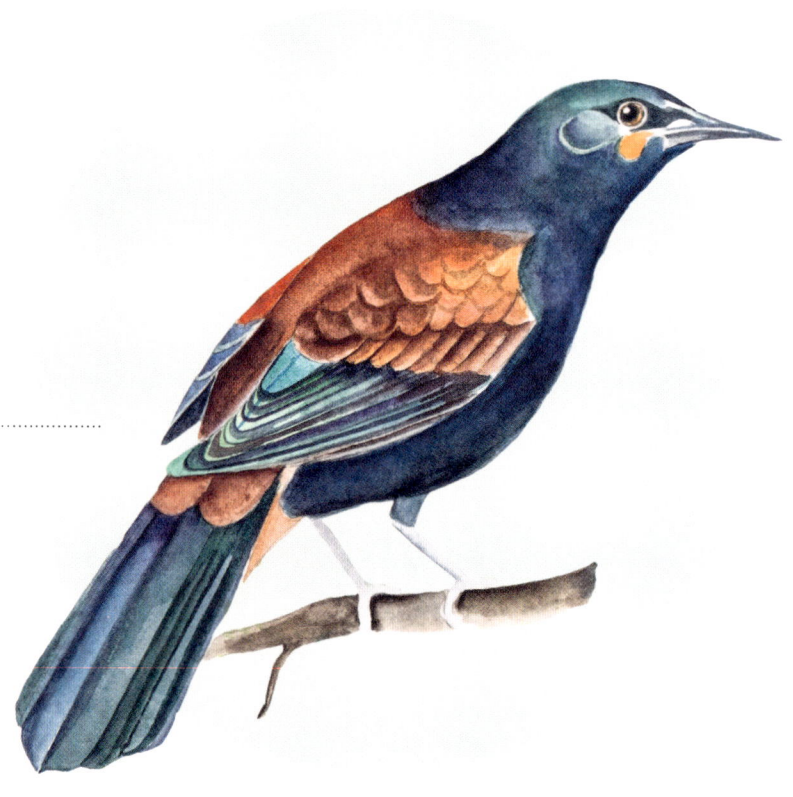

1.
NORTH ISLAND
SADDLEBACK

Philesturnus rufusater
Order Passeriformes

In a conservation success story, several of these threatened songbird species were relocated after threats to their habitat were identified and scientists found safer places for them to live.

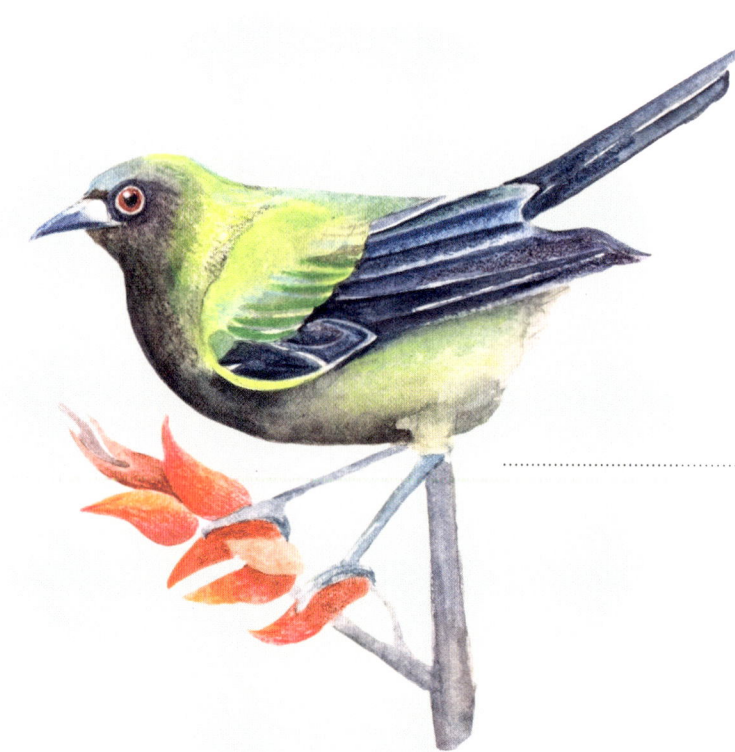

2.
NEW ZEALAND
BELLBIRD

Anthornis melanura
Order Passeriformes

These birds use their tubular, brush-tipped tongues to reach nectar inside flowers.

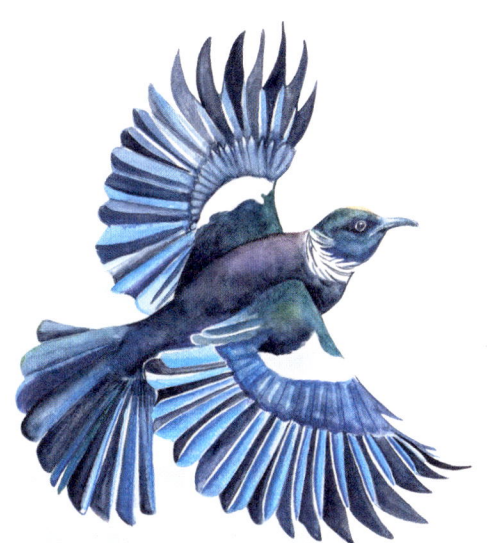

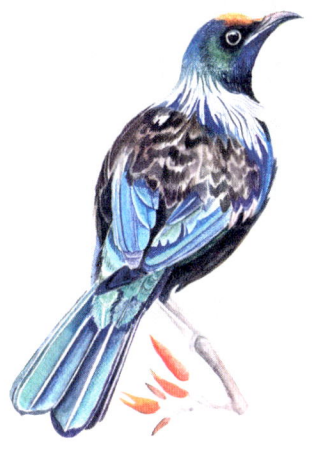

3. & 10.
TŪĪ

Prosthemadera novaeseelandiae
Order Passeriformes

Because they transfer pollen to various native plants, these nectar-feeders are an important part of the New Zealand ecosystem.

4.
NEW ZEALAND FANTAIL

Rhipidura fuliginosa
Order Passeriformes

These constantly moving songbirds often fly close to humans so that they can feed on the insects stirred up by people as they walk.

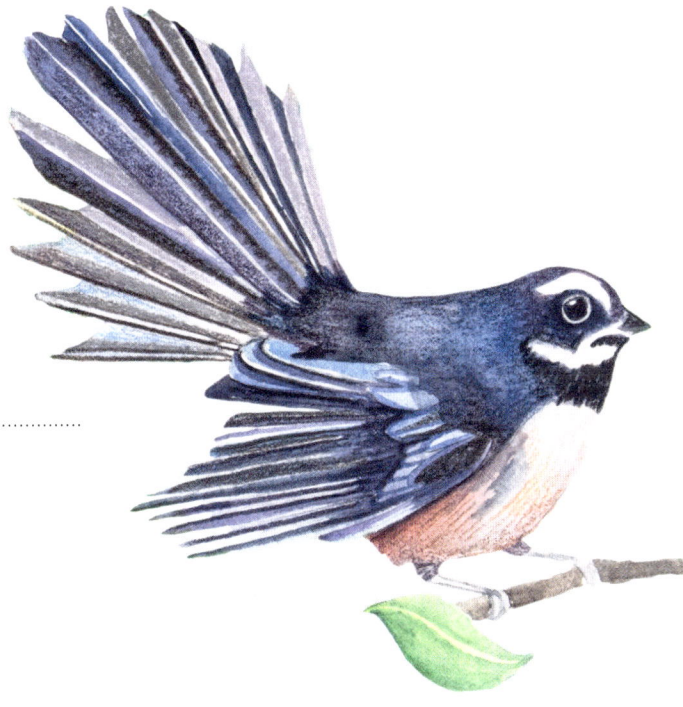

75

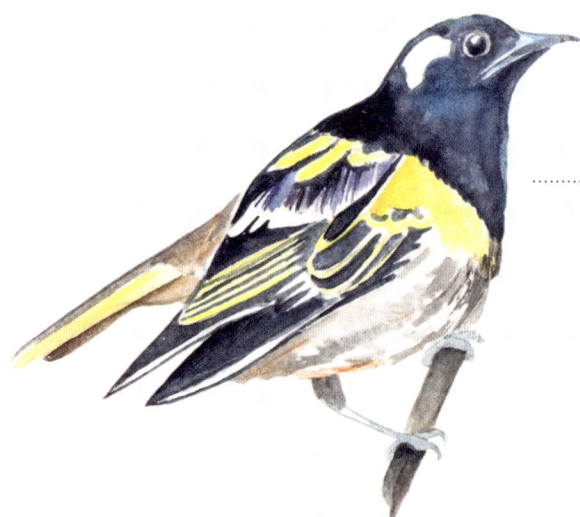

5.
STITCHBIRD (*HIHI*)

Notiomystis cincta
Order Passeriformes

One of New Zealand's rarest birds, the Stitchbird
is the only member of the family Notiomystidae.
Until recently, it was considered a member of
the honeyeaters, the family Meliphagidae.

6.
NEW ZEALAND
PIGEON (*KERERŪ*)

Hemiphaga novaeseelandiae
Order Columbiformes

Like many New Zealand birds, these pigeons are
vulnerable to introduced mammals such as possums
and rats, which eat the birds' eggs and nestlings.

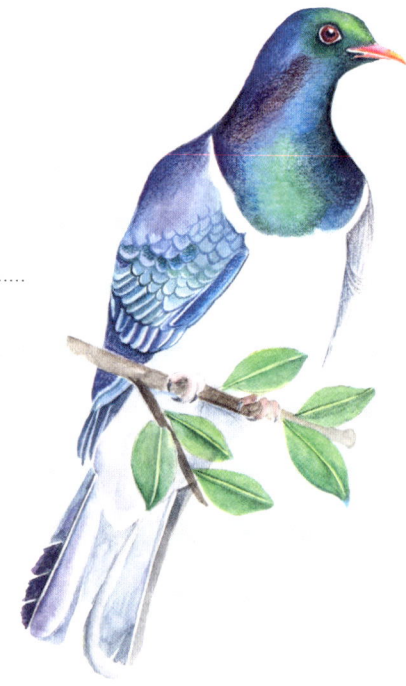

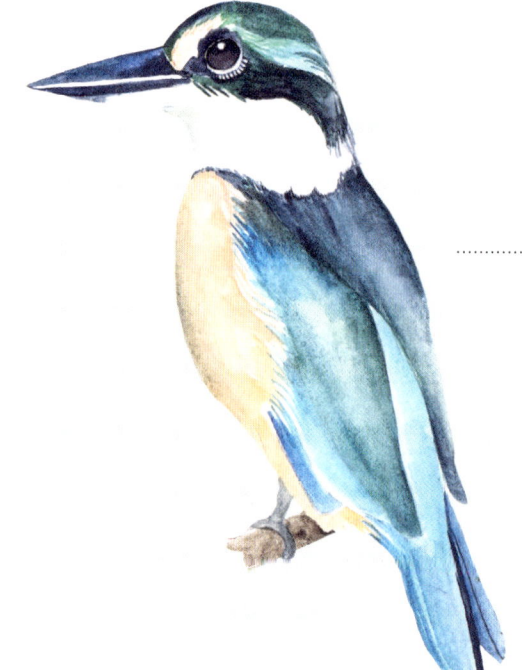

7.
SACRED KINGFISHER

Todiramphus sanctus
Order Coraciiformes

When nesting near water, Sacred Kingfishers
feed mainly on crustaceans, frogs, and
sometimes fish, but they also eat insects,
small rodents, and other small birds.

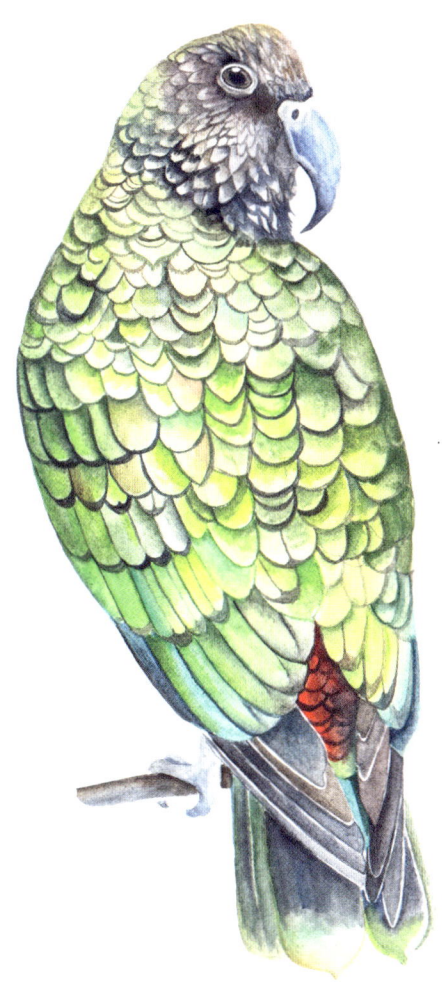

8.
KEA

Nestor notabilis
Order Psittaciformes

Controversy exists around the Kea, as the wild parrots use their powerful claws and beaks to pierce the skin of sheep and eat the fat when other food sources are unreliable. It was once common for farmers to kill the birds for bounty, but today the rare species is protected.

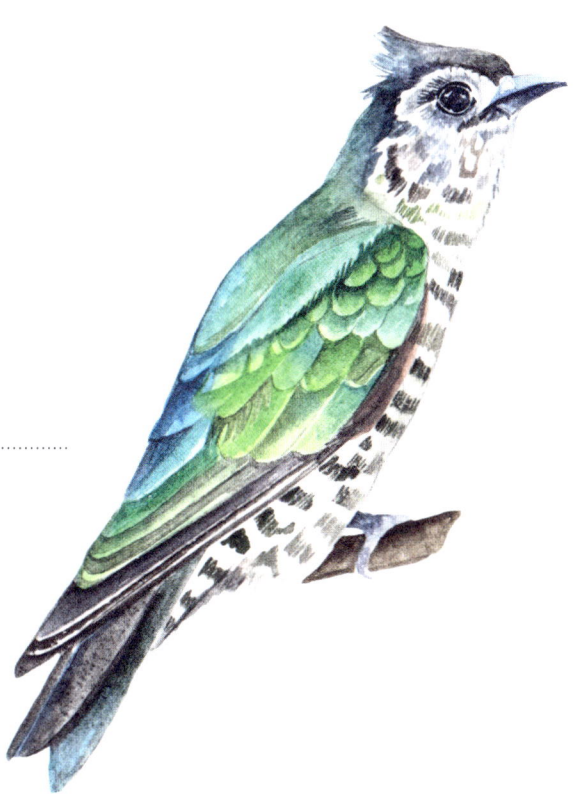

9.
SHINING
BRONZE-CUCKOO

Chrysococcyx lucidus
Order Cuculiformes

These birds practice brood parasitism—each female lays a single egg in the nest of another bird species and removes one of the host's eggs. After hatching, the cuckoo chicks are reared by the host parents.

BIRDS of AUSTRALIA'S NORTHERN TERRITORY

If you are afraid of venomous snakes, giant spiders, and fatally poisonous sea life, Australia is probably your worst nightmare. But if you love birds and beautiful backcountry, Australia is a paradise, and it may be worth putting your fears aside for a visit to see all the gorgeous birds this wonderful country has to offer. Much of the Australian landscape is quite dry, but it also has areas of dense rainforest and lush alpine mountain ranges. Birds from various parts of the continent are featured here, although most tend to be found in Australia's Northern Territory.

Many wild Australian birds are known to those who have never visited the continent because several species that are popular as caged birds are native to the Australian Outback. For example, the Northern Rosella (2), the Red-collared Lorikeet (7), and the Hooded Parrot (8) are all species that can be tamed and kept as pets. The Double-barred Finch (1) is also a popular pet bird.

The Spangled Drongo (5), a glossy black bird with startling red eyes and a dramatically forked tail, is a relatively common sight. In Australia, the word *drongo* is slang term for a foolish person—someone who can't seem to do anything right. Though this term was derived indirectly from the bird, it's not a bad reflection on this species! In the 1920s, a racehorse was named Drongo, after the bird, and despite his excellent breeding, the thoroughbred famously failed to produce a winning record and is responsible for the pejorative term used to this day.

A regular visitor to parks and backyards, the Magpie-lark (9) is also commonly referred to as a "peewee" and is sometimes confused with the Australian Magpie. These birds can sing in unison to intimidate rivals in their territory. Depictions of the Magpie-lark appear as symbols on various state and regional insignia.

1.	3.	5.	7.	9.
DOUBLE-BARRED FINCH	ROSE-CROWNED FRUIT DOVE	SPANGLED DRONGO	RED-COLLARED LORIKEET	MAGPIE-LARK
2.	4.	6.	8.	
NORTHERN ROSELLA	GOULDIAN FINCH	RED-HEADED MYZOMELA	HOODED PARROT	

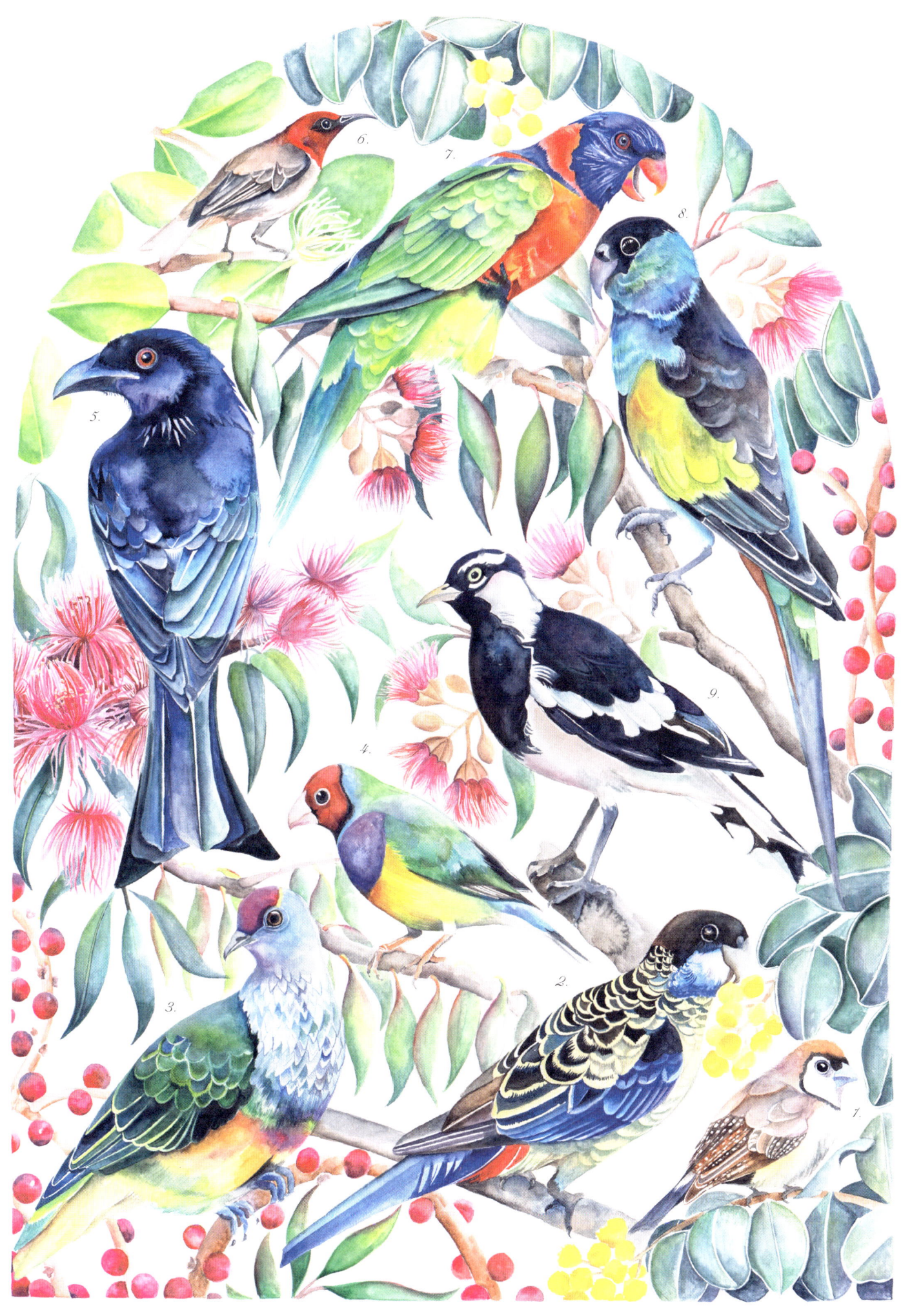

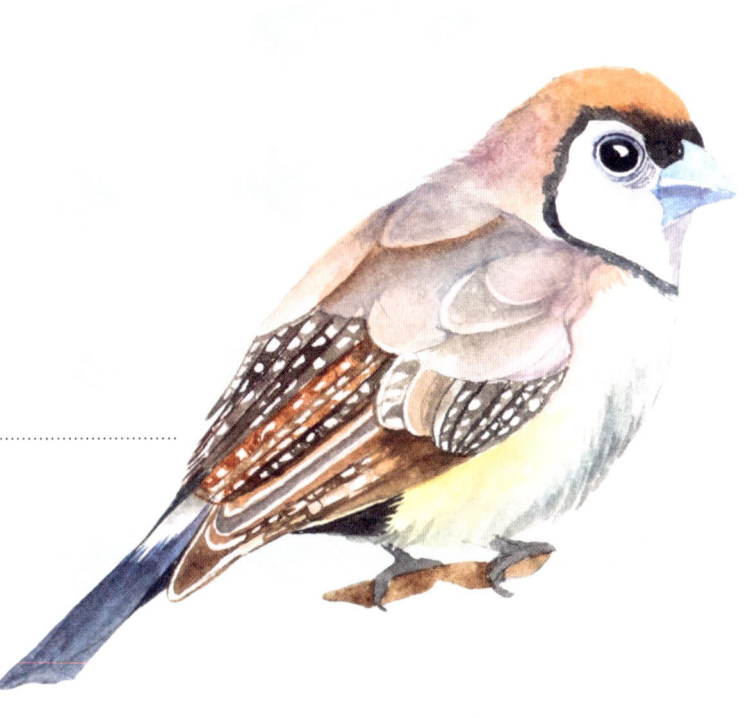

1.
DOUBLE-BARRED FINCH

Stizoptera bichenovii
Order Passeriformes

Because of its facial resemblance to a barn owl, this little finch is sometimes referred to as an "owl finch."

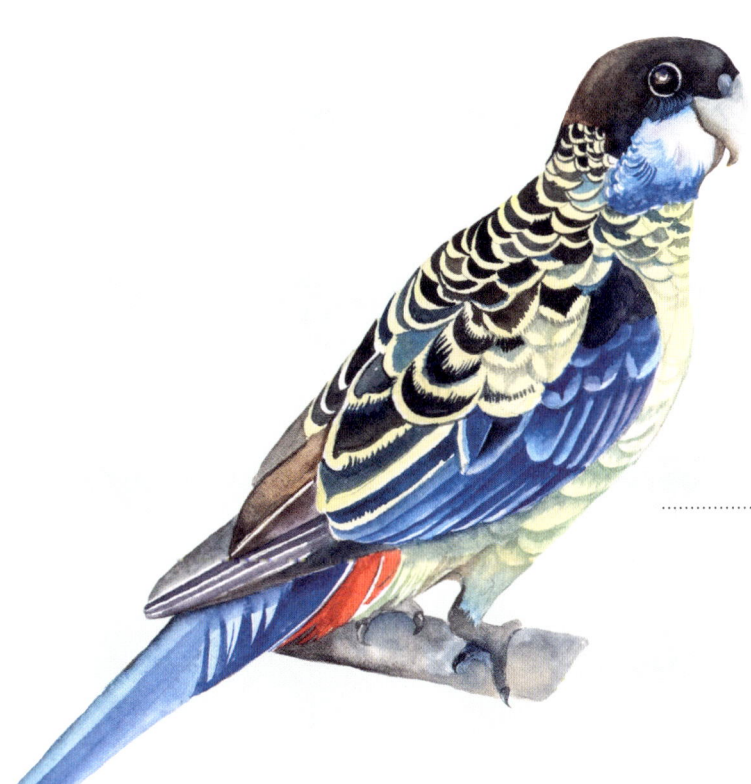

2.
NORTHERN ROSELLA

Platycercus venustus
Order Psittaciformes

Although native to Australia, this species was introduced to New Zealand in the early 1900s and has now naturalized there.

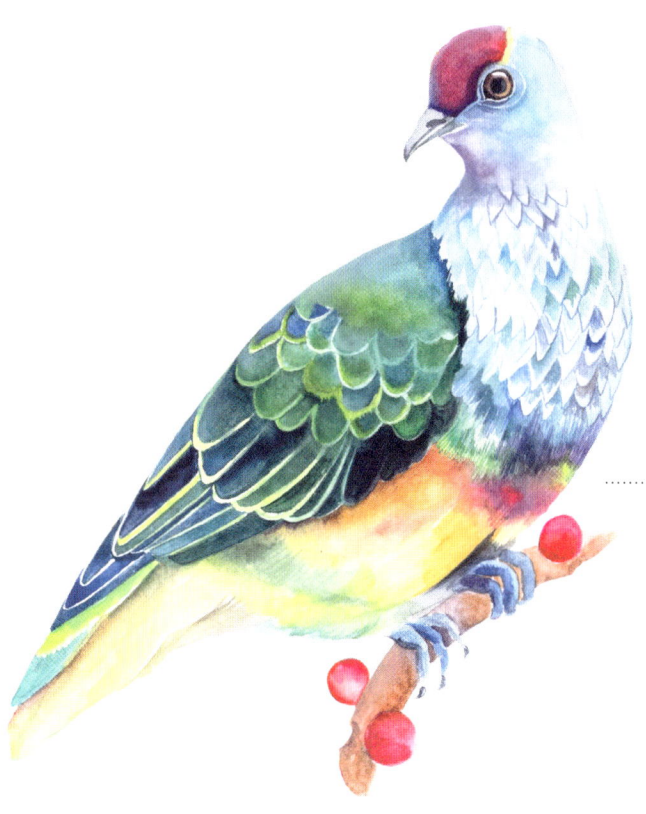

3.
ROSE-CROWNED FRUIT DOVE

Ptilinopus regina
Order Columbiformes

During breeding season, these doves perform a bowing display by tucking in their heads to show off their pink caps.

4.
GOULDIAN FINCH

Chloebia gouldiae
Order Passeriformes

Only the male Gouldian Finch sings. The females make simple chirps and calls.

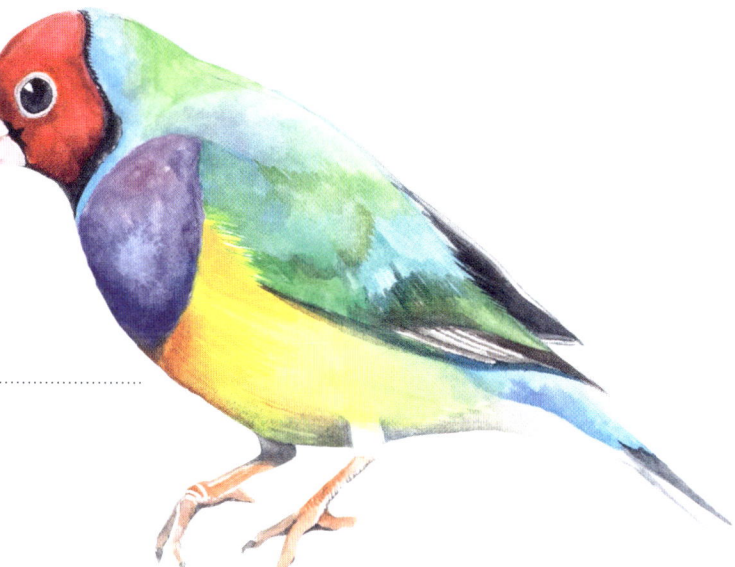

81

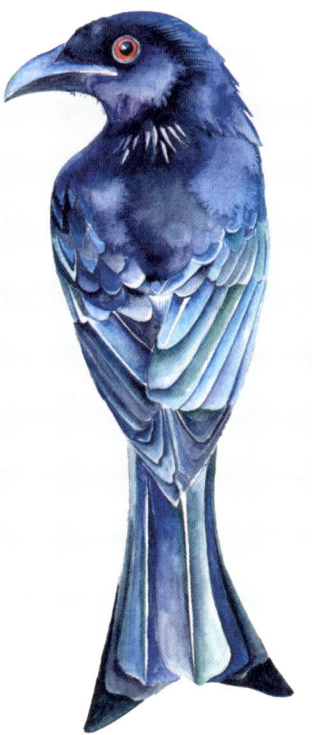

5.
SPANGLED DRONGO

Dicrurus bracteatus
Order Passeriformes

This songbird occasionally mimics
the calls of other birds.

6.
RED-HEADED MYZOMELA

Myzomela erythrocephala
Order Passeriformes

Typically found in coastal habitats, this bird mostly
inhabits mangroves and adjacent forests.

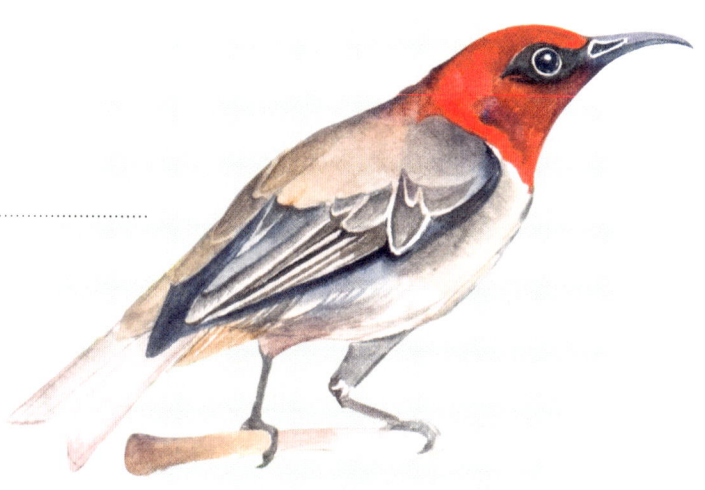

7.
RED-COLLARED LORIKEET

Trichoglossus rubritorquis
Order Psittaciformes

These birds are known to become inebriated
from consuming fermented nectar from fruit.
Affected birds can be seen clumsily running
and flying, and they may be particularly
friendly and uninhibited around humans.

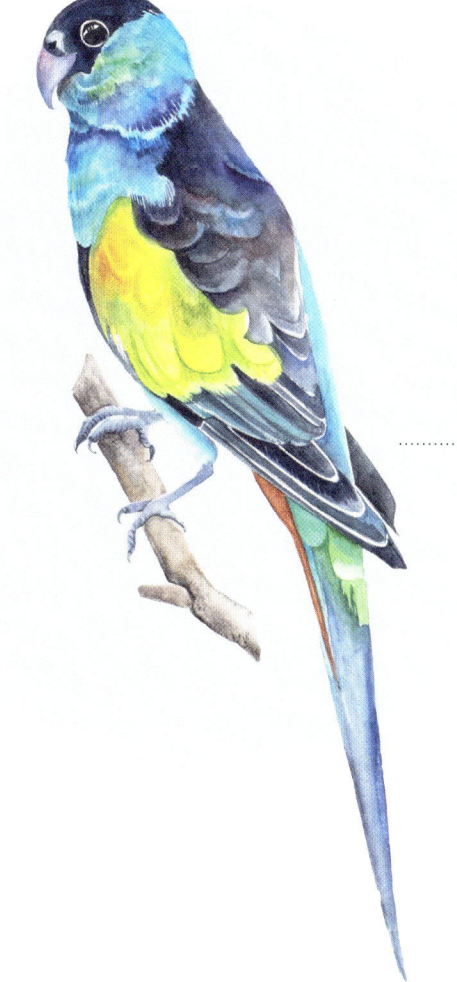

8.
HOODED PARROT

Psephotellus dissimilis
Order Psittaciformes

This species often digs tunnel nests in termite
mounds, where they lay four to six eggs.

9.
MAGPIE-LARK

Grallina cyanoleuca
Order Passeriformes

This is one of a small number of birds worldwide
that sing coordinated duets, with breeding pairs
singing together to mark their territory.

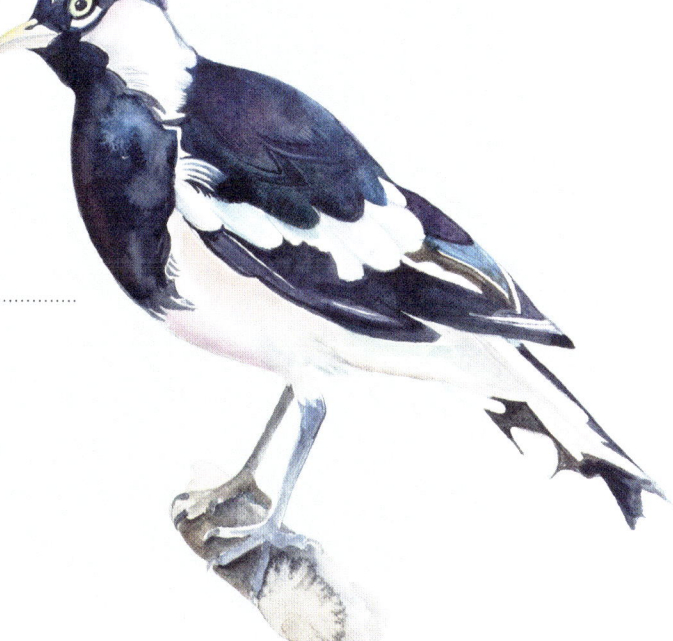

83

BIRDS OF AUSTRALIA'S NORTHERN TERRITORY ⋯⋯⋯⋯⋯⋯⋯ ✴

BACKYARD BIRDS of the HAWAIIAN ISLANDS

The Hawaiian Islands of the Pacific Ocean are world renowned for the scope of their natural beauty. Hawaii is one of the most ecologically diverse places in the world. The islands are home to a vibrant tapestry of lush jungle flora, emerald-green peaks, cascading waterfalls, black sand beaches, and impressive, imposing volcanoes. Although home to many notable native and endemic species, the Hawaiian Islands are unique in their abundance and variety of successfully established introduced and invasive birds. A bird enthusiast could encounter species in Hawaii that originated on nearly every continent in the world.

One common introduced bird is the Red Junglefowl (1), which is sometimes referred to as the "original chicken," because nearly all domesticated chickens are descended from the wild junglefowl. Native to Southeast Asia, this species was introduced by early Polynesian settlers as a source of meat, and the birds have since established themselves and now breed with domestic chickens, making them and their relatives ubiquitous throughout the islands.

Hawaii is also home to many endemic birds in the honeycreeper family (Fringillidae). Their long, slender, sickle-shaped bills and brush-tipped tongues aid them in reaching the nectar within flowers. Unlike hummingbirds, which hover above flowers beating their wings rapidly to stay aloft, honeycreepers perch atop flowers and pierce them with their long beaks to reach the nectar within. Sadly, many species of honeycreepers have been driven to extinction by habitat destruction and urban development, but a few, such as the O'ahu 'Amakihi (2), the 'Ākohekohe (3), and the 'I'iwi (4), may be seen by lucky birdwatchers.

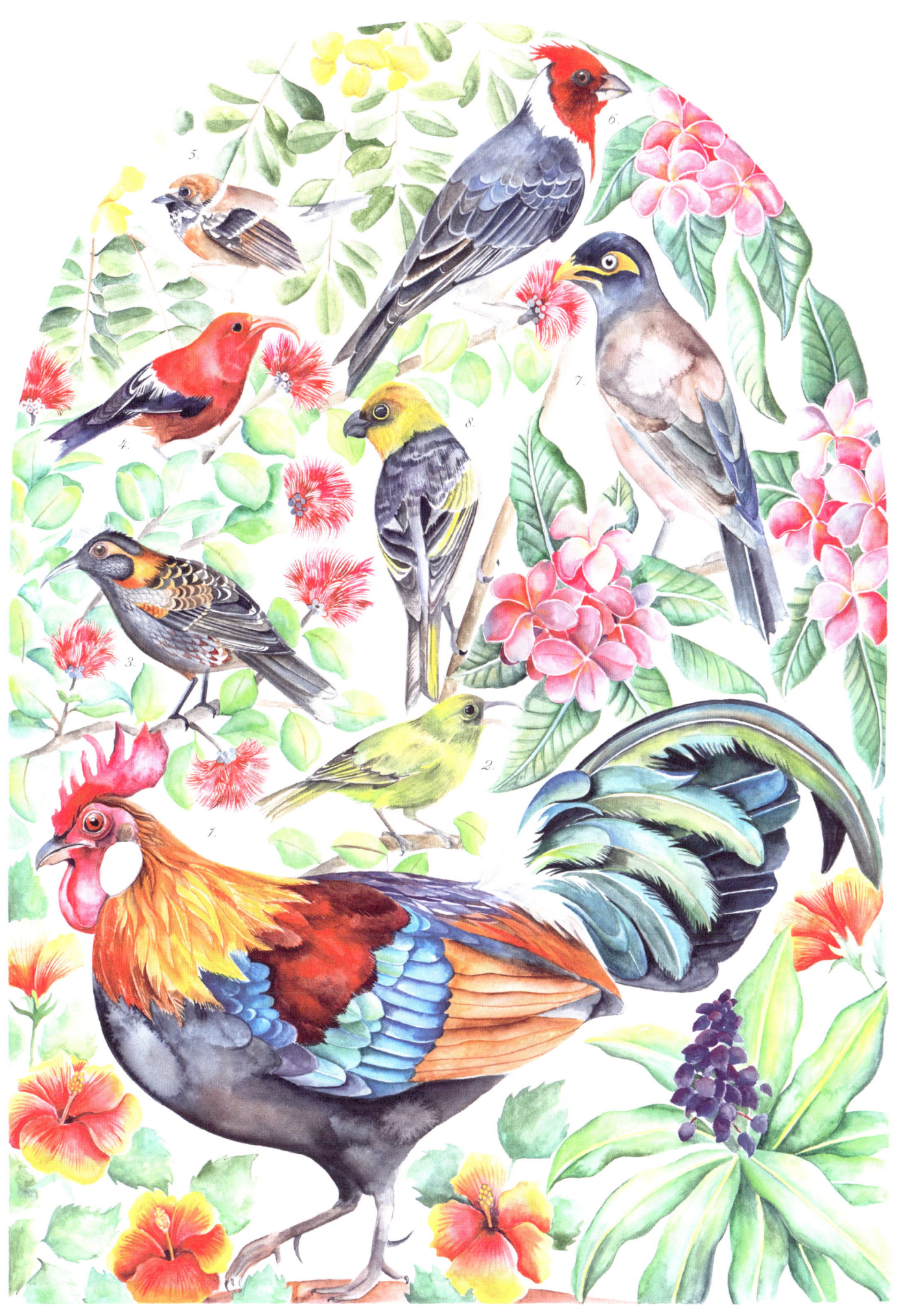

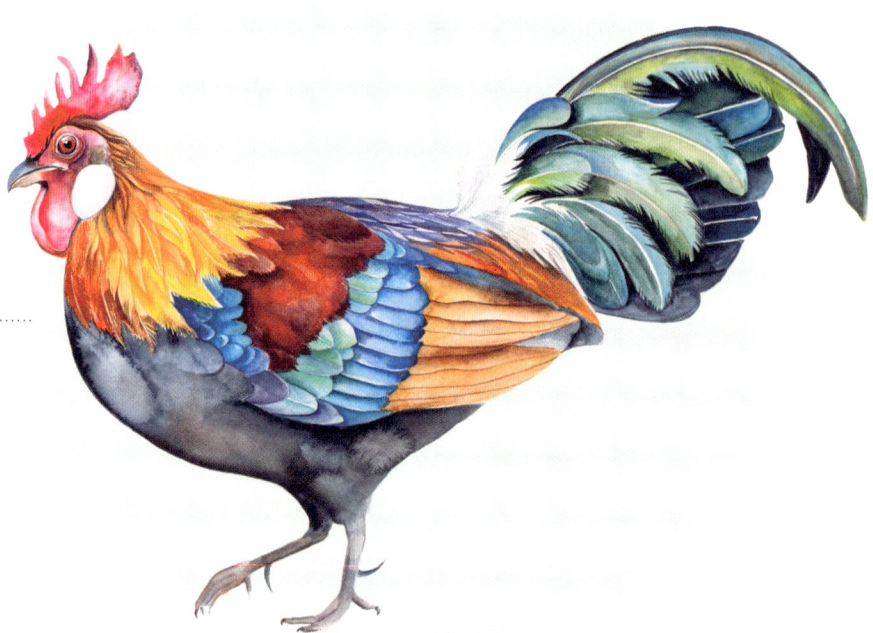

1.
RED JUNGLEFOWL

Gallus gallus
Order Galliformes

In addition to eating the meat of these birds, early Polynesian settlers used their feathers to adorn royal and ceremonial garments.

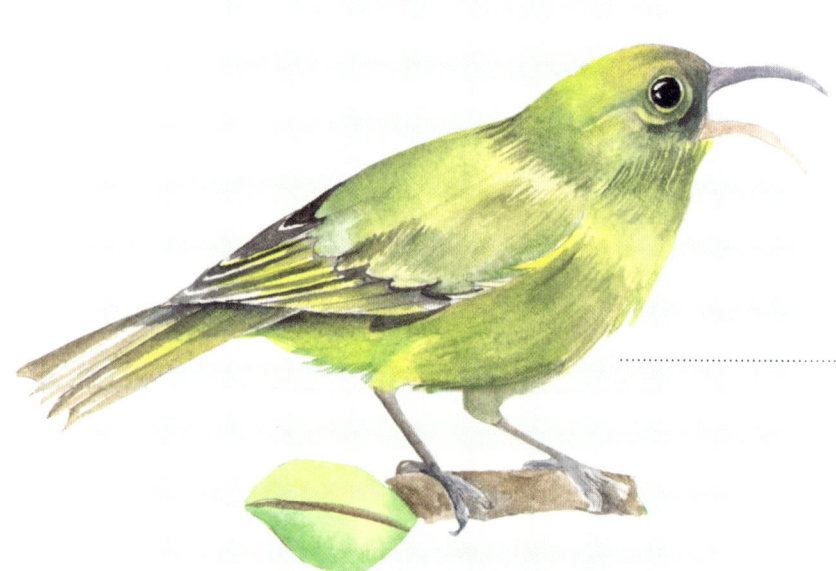

2.
OʻAHU ʻAMAKIHI

Chlorodrepanis flava
Order Passeriformes

This tiny honeycreeper is endemic to the island of Oahu.

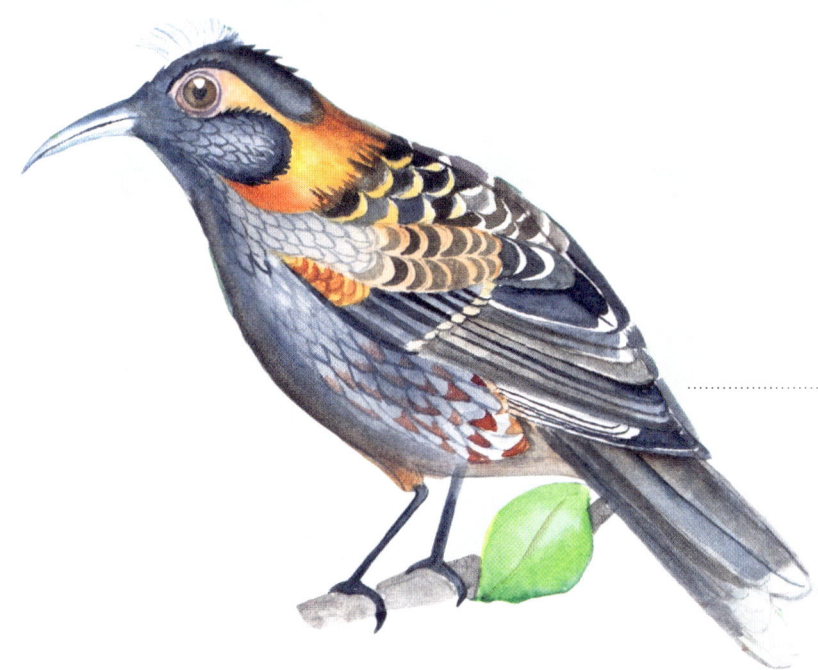

3.
ʻĀKOHEKOHE (CRESTED HONEYCREEPER)

Palmeria dolei
Order Passeriformes

At about 7 inches long, this critically endangered species is one of the largest of all honeycreepers.

4.
ʻIʻIWI (SCARLET HONEYCREEPER)

Drepanis coccinea
Order Passeriformes

This endangered species is one of the most beautiful honeycreepers and a widely recognizable symbol of the Hawaiian Islands.

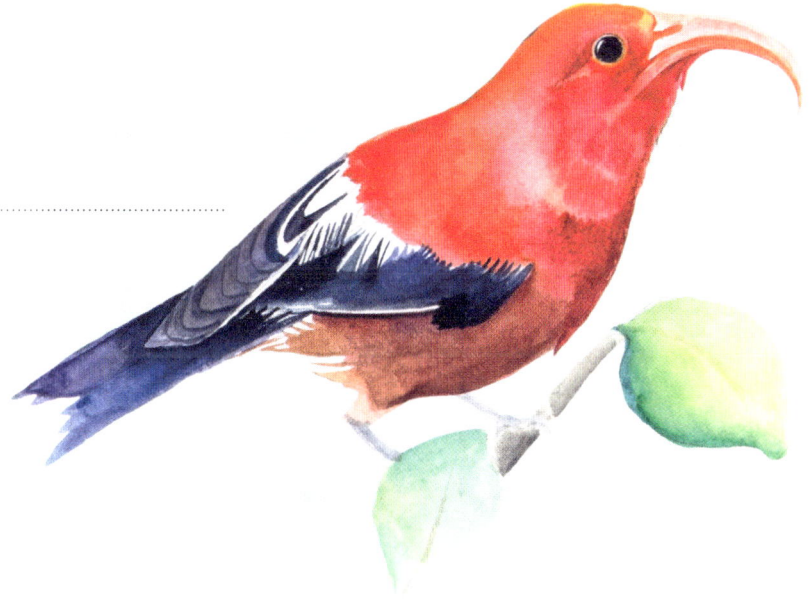

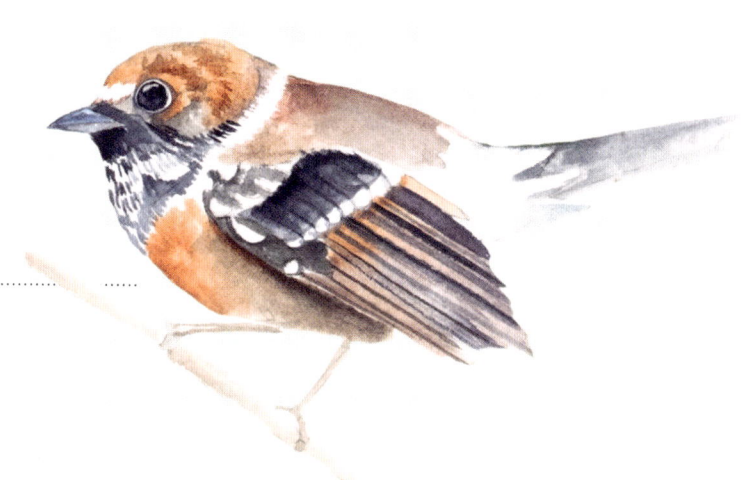

5.
HAWAI'I 'ELEPAIO

Chasiempis sandwichensis
Order Passeriformes

The common name of this little flycatcher is onomatopoeic for the male's shrill song.

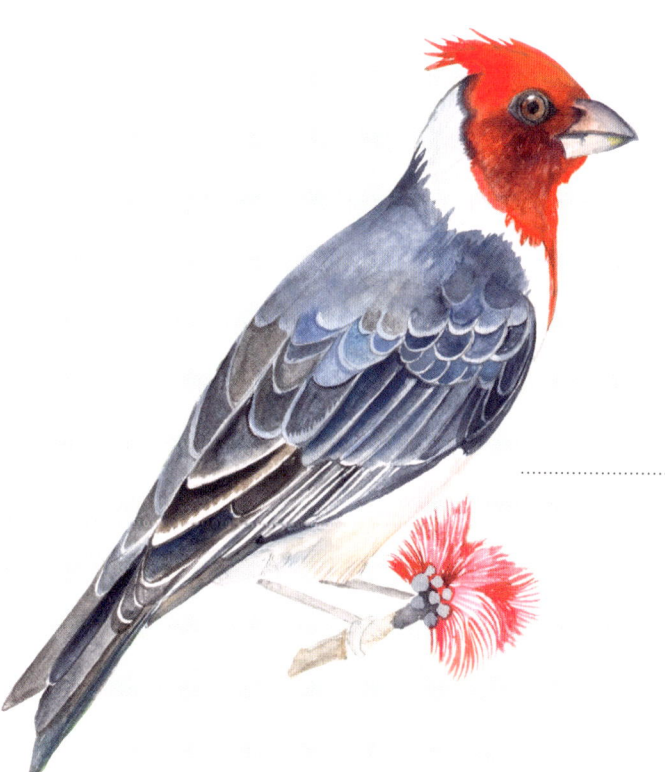

6.
YELLOW-BILLED
CARDINAL

Paroaria capitata
Order Passeriformes

Despite its common name, this South American native bird is not closely related to the bright red Northern Cardinal common in the Central and Eastern United States.

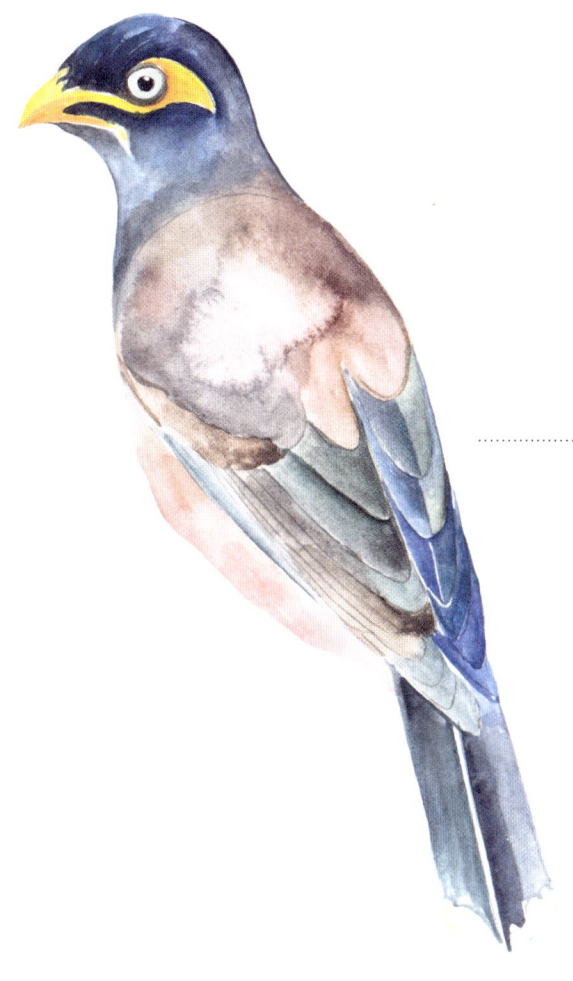

7.
COMMON MYNA

Acridotheres tristis
Order Passeriformes

Originally from Asia, this adaptable bird has been introduced to many other parts of the world and is now considered one of the world's most invasive species.

8.
YELLOW-FRONTED CANARY

Crithagra mozambica
Order Passeriformes

Native to Africa, this species was introduced to the Hawaiian Islands in the 1960s and has established itself on Oahu and the island of Hawaii.

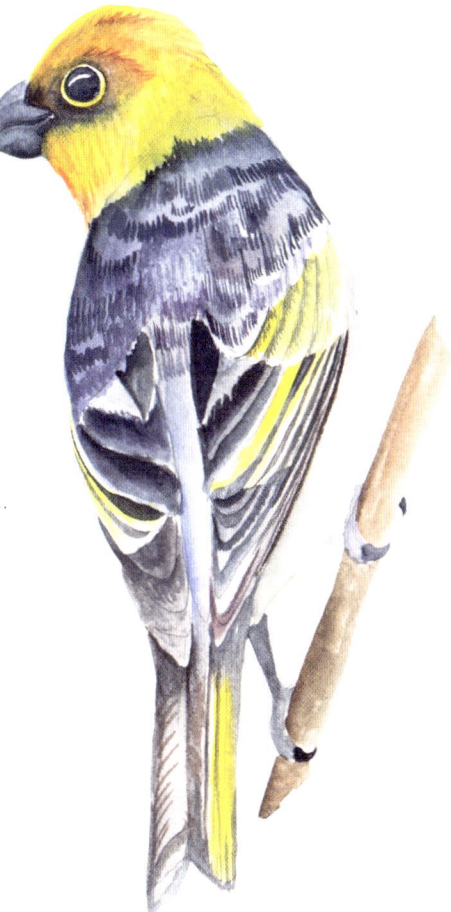

BIRDS
of
PAPUA
NEW GUINEA

This island country in the southwestern Pacific Ocean is renowned as one of the foremost birdwatching destinations in the world, boasting one of the highest concentrations of endemic species on Earth. The country is mainly mountainous, with many active volcanoes and the world's third largest rainforest.

The blazing red-and-orange plumage of the Flame Bowerbird (3) is only a small part of what makes this native bird special. The male performs a courtship dance so remarkable that it has inspired various filmmakers and documentarians to record the exceptional display. Male bowerbirds build elaborate nests, or bowers, composed of twigs that are bent and configured to create a kind of portal. When a prospective female mate approaches the bower, the male begins its intricate dance with wings extended and chest and feathers puffed.

Papua New Guinea is also host to several of the incredible varieties of birds-of-paradise, and lucky visitors may catch a glimpse of the Brown Sicklebill (7), a family member identifiable by its graceful, flowing tail. The slightly less elegant Pesquet's Parrot (4) is a bizarre looking parrot with a bald head and a curved, vulturine beak.

The gorgeous Victorian Crowned Pigeon (5), the world's largest pigeon, is often kept in many zoos and aviaries throughout the world. It's no mystery as to why this species would be a desirable bird in captivity, with its eye-catching, lace-like fan of white-and-purple feathers. Sadly, this unique and appealing feathery crested bird is also sought out by hunters, who illegally trap and sell the birds.

1.
HOODED PITOHUI

2.
BELFORD'S
MELIDECTES

3.
FLAME BOWERBIRD

4.
PESQUET'S PARROT

5.
VICTORIA CROWNED-
PIGEON

6.
PALM COCKATOO

7.
BROWN SICKLEBILL

8.
PAPUAN KING-PARROT

9
PAPUAN DWARF-
KINGFISHER

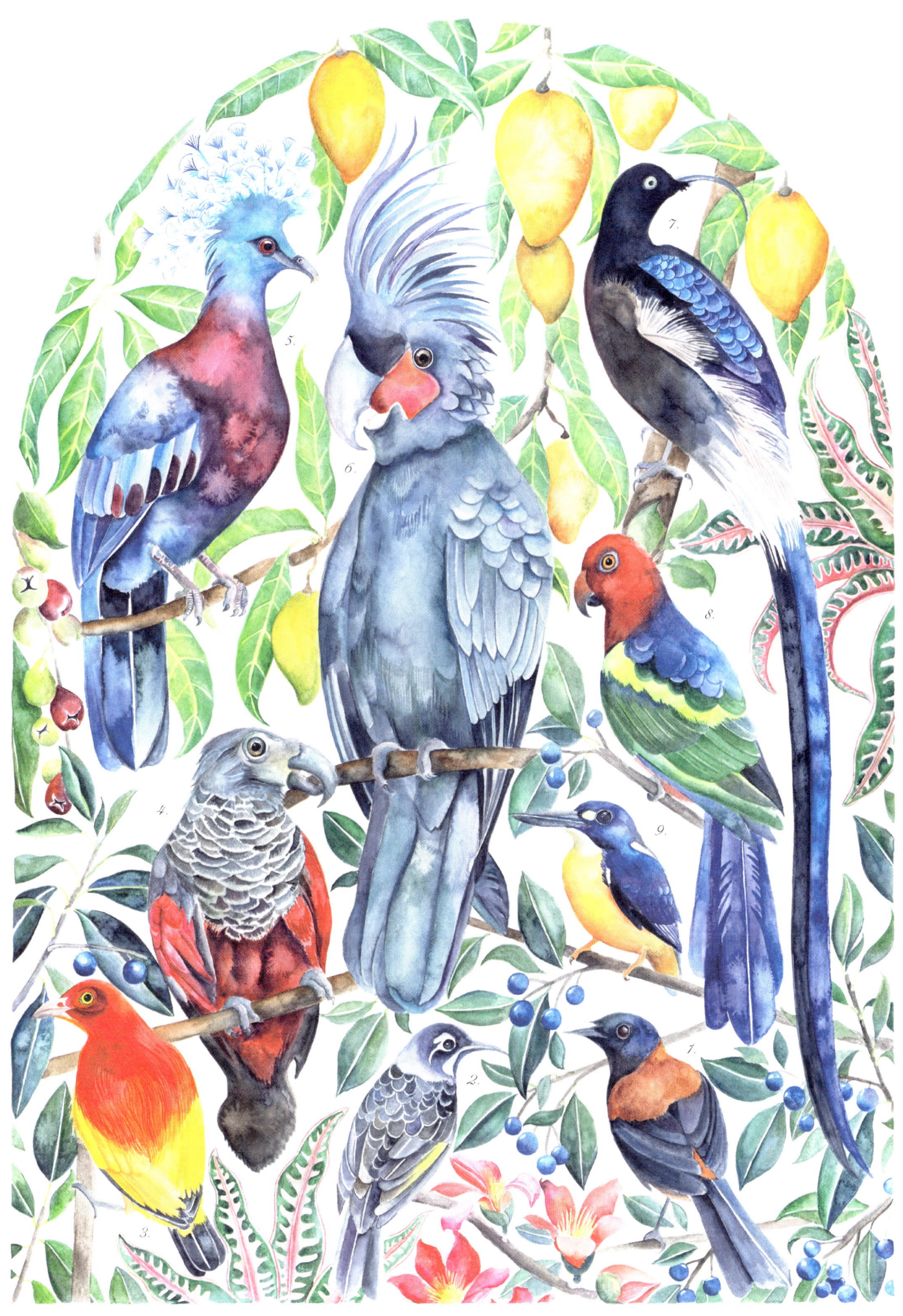

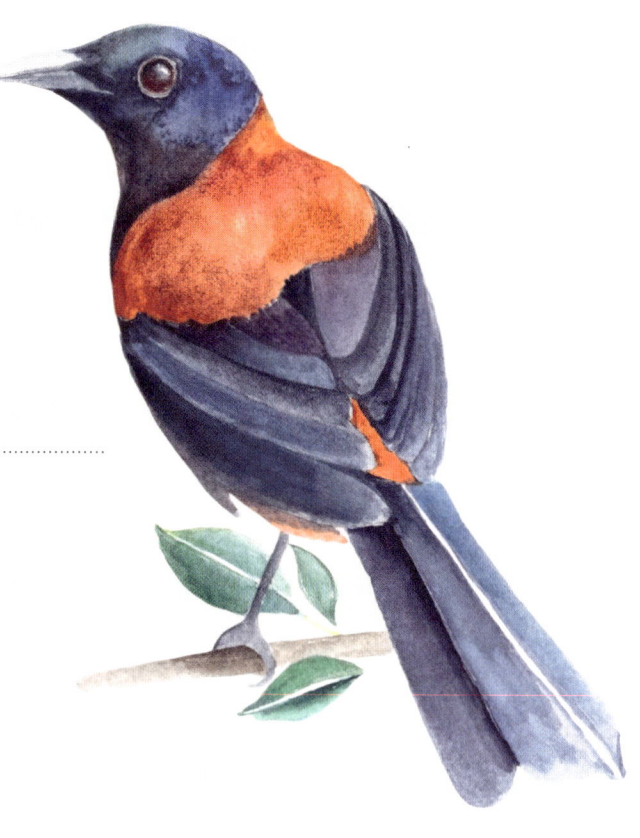

1.
HOODED PITOHUI

Pitohui dichrous
Order Passeriformes

This medium-sized species is most easily identified
by its distinctive *pi-to-hui* call and a series
of low warbles, wolf-whistles, and clicks.

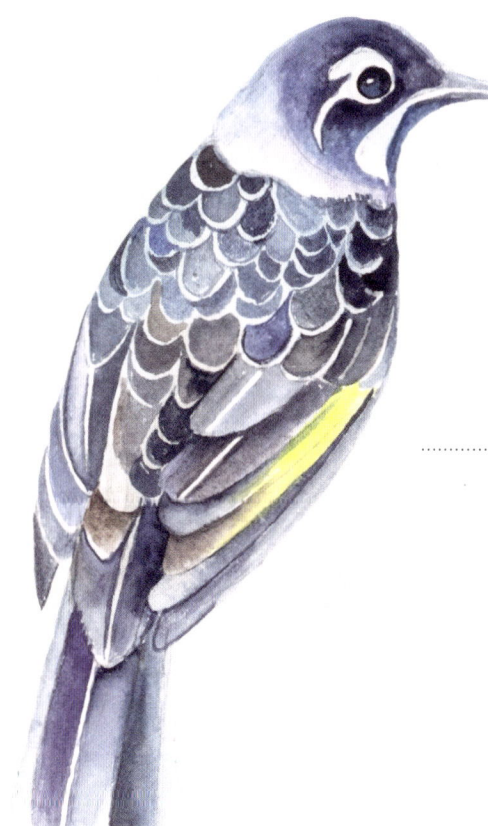

2.
BELFORD'S
MELIDECTES

Melidectes belfordi
Order Passeriformes

These birds feed mainly on insects and nectar
from flowering trees. They hop throughout
the forest canopy, probing flowers, mosses,
and epiphytes in search of nourishment.

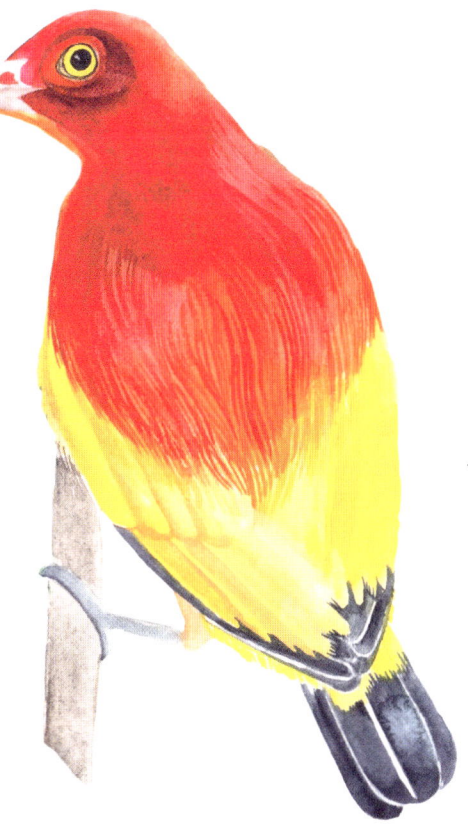

3.
FLAME BOWERBIRD

Sericulus ardens
Order Passeriformes

During the male bowerbird's elaborate courtship
dance, he offers berries and other colorful
objects to the female to entice her.

4.
PESQUET'S PARROT

Psittrichas fulgidus
Order Psittaciformes

This parrot's diet consists almost entirely of
fruit, specifically figs, and it is believed that its
bald head is an evolutionary adaptation to avoid
sticky feathers when gorging on the juicy fruit.

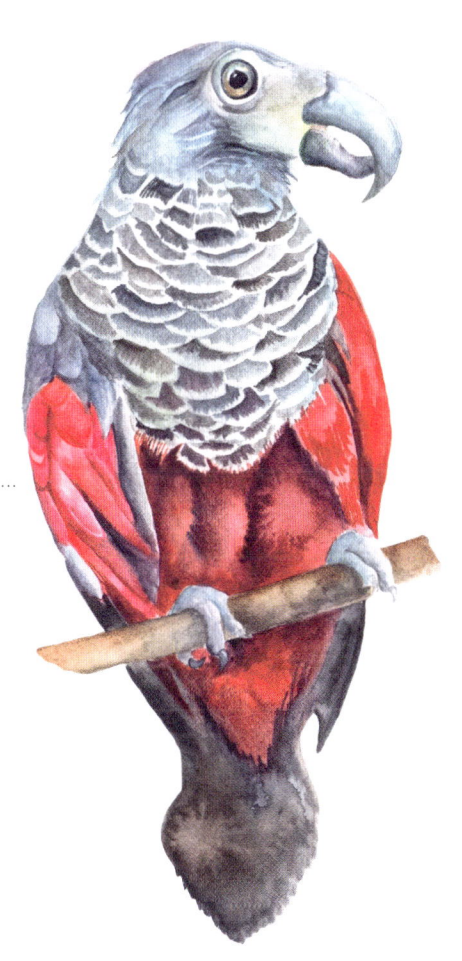

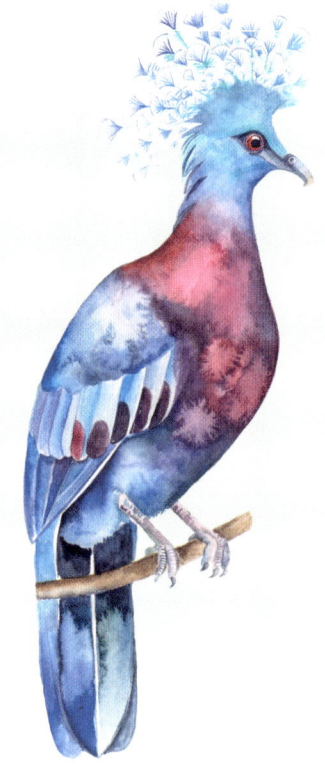

5.
VICTORIA
CROWNED-PIGEON

Goura victoria
Order Columbiformes

Flocks of these pigeons forage on
the forest floor for fallen fruit.

6.
PALM COCKATOO

Probosciger aterrimus
Order Psittaciformes

The bare red patches on this cockatoo's
cheeks can change to a pinkish beige
when the bird is alarmed or excited.

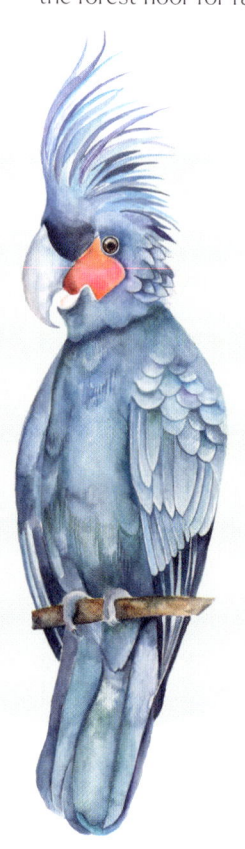

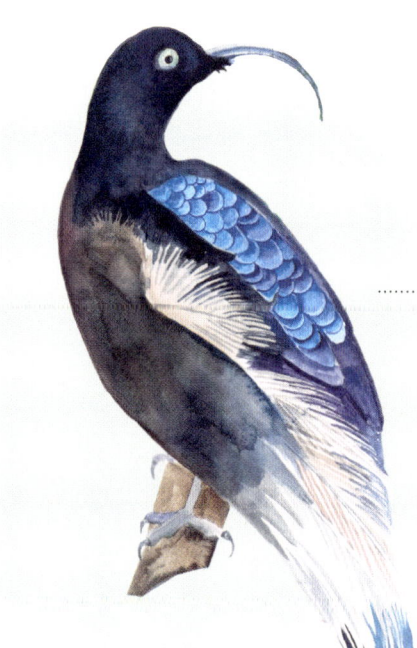

7.
BROWN SICKLEBILL

Epimachus meyeri
Order Passeriformes

This bird-of-paradise's genus name is an
ancient Greek word meaning "equipped
for battle" and was perhaps assigned to
the bird because of its sword-like bill.

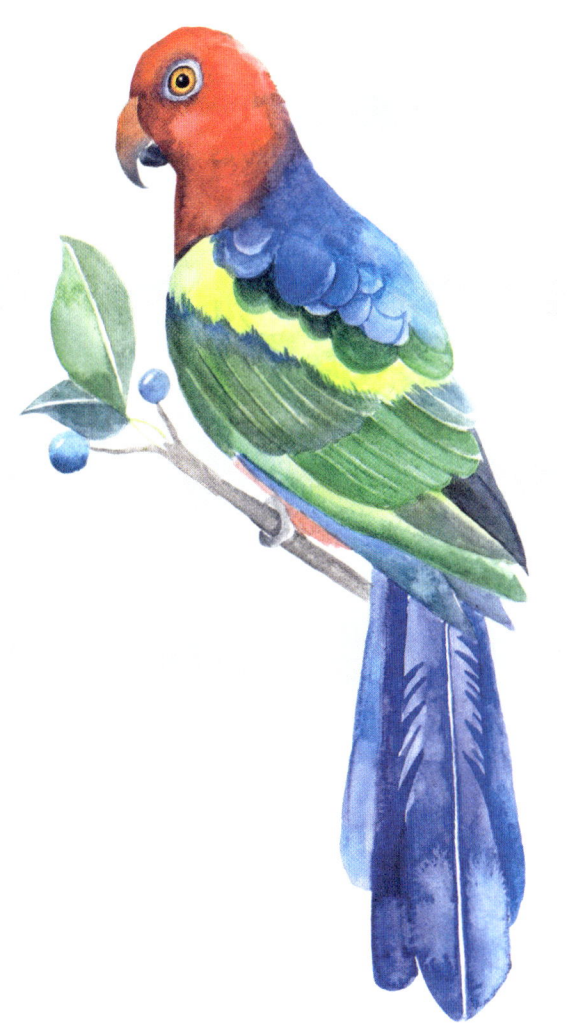

8.
PAPUAN KING-PARROT

Alisterus chloropterus
Order Psittaciformes

Though the male and female of this species
look somewhat alike, the male parrot has
slightly deeper and brighter red feathers.

9.
PAPUAN
DWARF-KINGFISHER

Ceyx solitarius
Order Coraciiformes

This deep-blue bird is one of the smallest of all
kingfishers and is extremely difficult to spot.

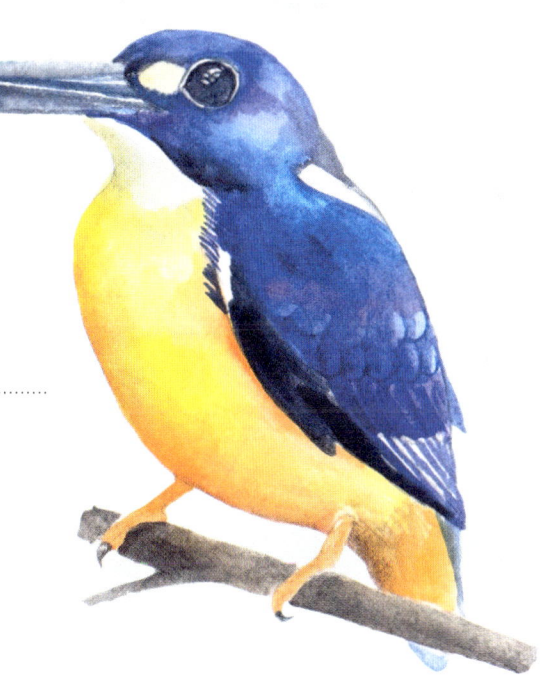

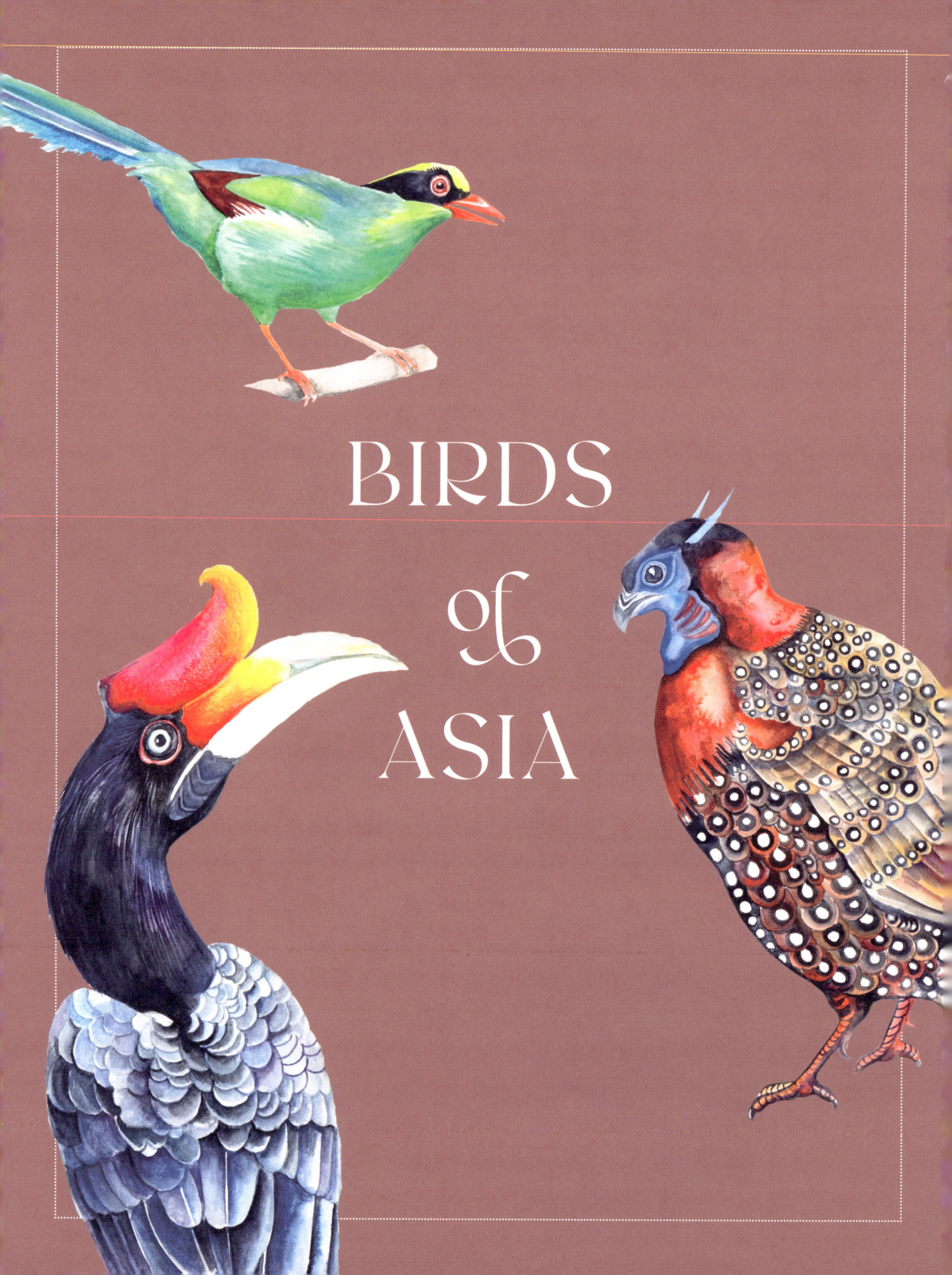

BIRDS

of

ASIA

The largest continent in the world has a correspondingly vast and varied selection of bird life. We will visit some of the most striking birds from across this huge continent and its variety of climates and geography.

Look for the slender and elegant Red-crowned Crane, iconic in myths and legends of various Asian cultures. Be sure not to miss the sweeping tail and brilliant fiery colors of the Golden Pheasant, believed by some to be the real-life inspiration for the mythical phoenix. You may also encounter one of several species of bird-of-paradise, which was once believed to fly without wings through the heavens.

BIRDS of CHINESE GARDENS

Birds appear in Chinese folklore as both omens of good fortune and harbingers of dark tidings. They flutter in and out of Chinese art as part of various cultural motifs. The Red-billed Blue-Magpie (2, 3) is a common bird in China and throughout Asia. Although they are related and similar in size to the Eurasian Magpie, the fluttering, striped tail of the Red-billed Blue-Magpie sets them apart. Native to forests throughout Asia, these raucous birds have adapted to live in urban spaces and are commonly seen in cities and cultivated gardens. The largest bird featured here is the Golden Pheasant (4), sometimes called the Rainbow Pheasant or Chinese Pheasant. Although native to the forests of China, this pheasant's kaleidoscopic plumage and elegant sweeping tail make it look almost otherworldly. It is no wonder that many believe it is related to the mythical firebird, the phoenix.

One of the most beautiful ducks in the world, the colorful Mandarin Duck (1) is native to China. They are often depicted in traditional Chinese art and pottery. Although they do not mate for life, they feature prominently in motifs and symbols of marriage and fidelity and are often pictured in wedding gifts and traditions to offer tidings of a faithful and enduring union. Males are very ornate and colorful, while females look entirely different, with grayish plumage, striped and dotted in white. Pairs are sometimes referenced as symbols of an odd couple. The Chinese word for this duck, *yuānyāng*, refers to the male (*yuān*) and female (*yāng*) of the species.

One Chinese folk tale recounts a young maiden, the daughter of a wealthy and important man, who slips and falls into a pond. Her father's gardener quickly plunges into the pond to save the young girl from drowning. The father sees the incident from a distance and mistakenly thinks the gardener is trying to assault his beloved daughter. He has the man imprisoned for his perceived insolence, but the daughter visits her savior in his cell and presents him with a beautiful technicolor robe. The father, upon discovering this gift, has the man thrown into the pond and drowned. Tragically, when his daughter learns of the fate of her savior, she too drowns herself in the pond. From that point forth, a pair of mandarin ducks were seen gliding across the pond, never far from one another—a reminder of the ill-fated lovers.

1.
MANDARIN DUCK

2. & 3.
RED-BILLED
BLUE-MAGPIE

4.
GOLDEN PHEASANT

5.
YELLOW-BILLED
GROSBEAK

6.
BLACK-NAPED ORIOLE

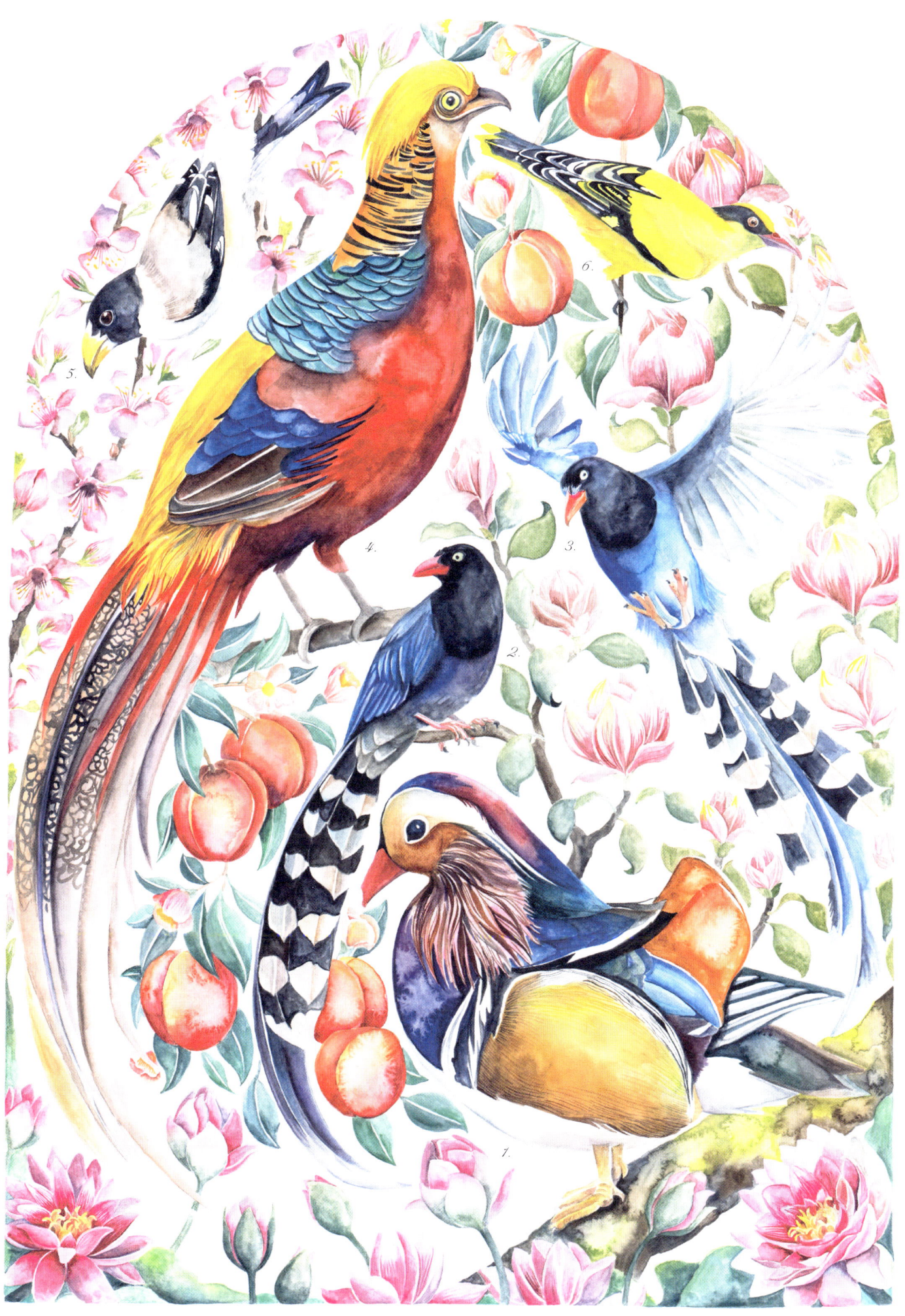

1.

MANDARIN DUCK

Aix galericulata
Order Anseriformes

These shy ducks tend to prefer smaller bodies of water such as ponds and shallow marshes, where they can easily find cover in dense neighboring woodlands.

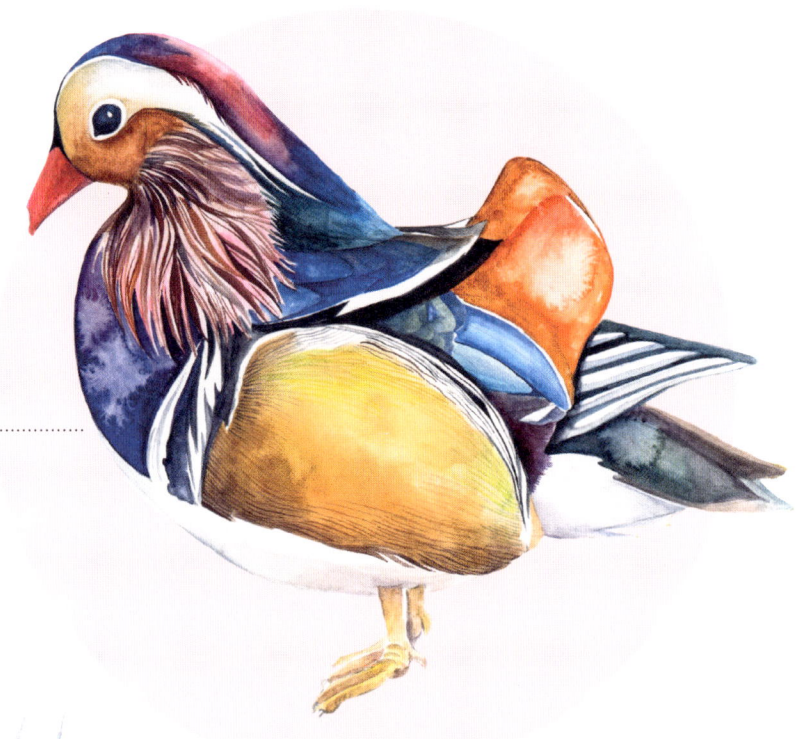

2. & 3.

RED-BILLED
BLUE-MAGPIE

Urocissa erythroryncha
Order Passeriformes

These magpies are highly territorial and aggressive and will fiercely defend their territory from perceived threats. They are also known to pillage the nests of other birds to feed on their eggs and hatchlings.

4.
GOLDEN PHEASANT

Chrysolophus pictus
Order Galliformes

The striking males of this species are incredibly colorful, with scarlet, orange, bright blue, and yellow plumage; the females are far more monochromatic.

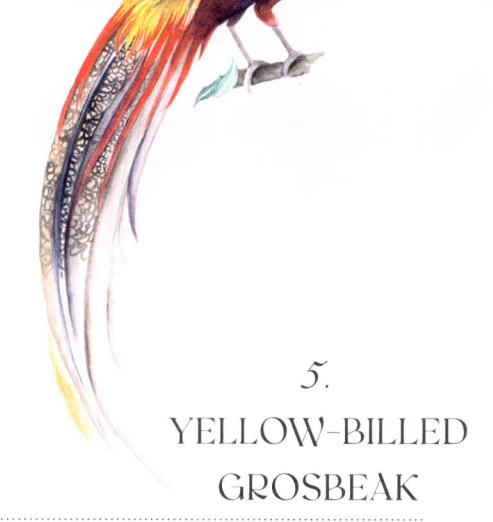

5.
YELLOW-BILLED GROSBEAK

Eophona migratoria
Order Passeriformes

Sometimes called Chinese Grosbeaks, these birds do not fear humans and frequent gardens, orchards, and farmland.

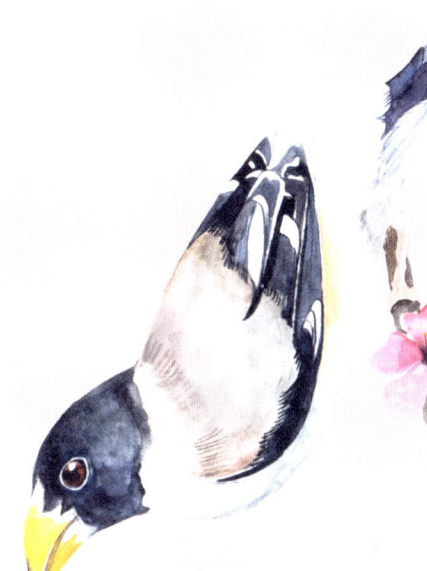

6.
BLACK-NAPED ORIOLE

Oriolus chinensis
Order Passeriformes

This brilliantly yellow bird is distinguishable by the black "bandit mask" across its eyes to its nape.

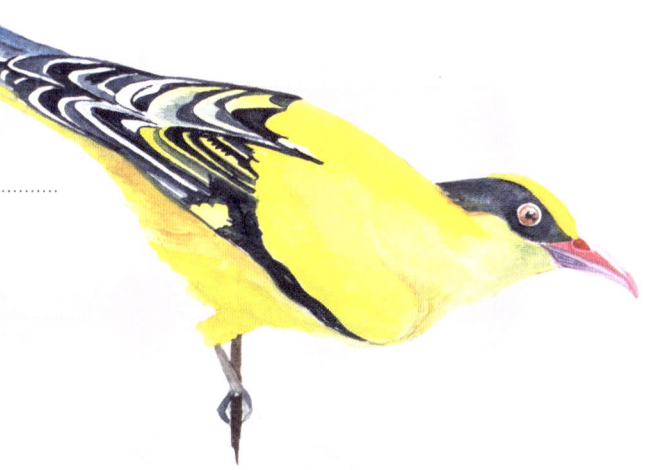

BIRDS of the HIMALAYAS

The immense Himalayan mountain range includes some of the highest peaks on Earth, the most famous of which is Mount Everest. Frequented by mountaineers seeking a thrilling view from the world's highest summits, this region is also rich with nature, including some truly extraordinary birds.

In the same family as the peacock and pheasant (Phasianidae), and just as beautiful, are the Blyth's Tragopan (1), the Himalayan Monal (3), and the Western Tragopan (8). The Himalayan Monal is perhaps the most iridescent and radiantly colored pheasant in the world. The national bird of Nepal, which is often mentioned in works of music and literature, the exquisite Himalayan Monal lives high in the mountains and is a rare but thrilling sight. They are sexually dimorphic (males and females do not look alike), and the brilliant metallic-feathered bird pictured here is the male of the species. The female is a muted brown and much less remarkable looking than the male. In mating season, the males perform an elaborate courtship dance, in which they display their magnificent plumage to a prospective mate.

1.
BLYTH'S TRAGOPAN

2.
BLUE-FRONTED REDSTART

3.
HIMALAYAN MONAL

4.
HIMALAYAN CUTIA

5.
GREAT BARBET

6.
VERDITER FLYCATCHER

7.
HIMALAYAN WHITE-BROWED ROSEFINCH

8.
WESTERN TRAGOPAN

9.
HIMALAYAN BLUETAIL

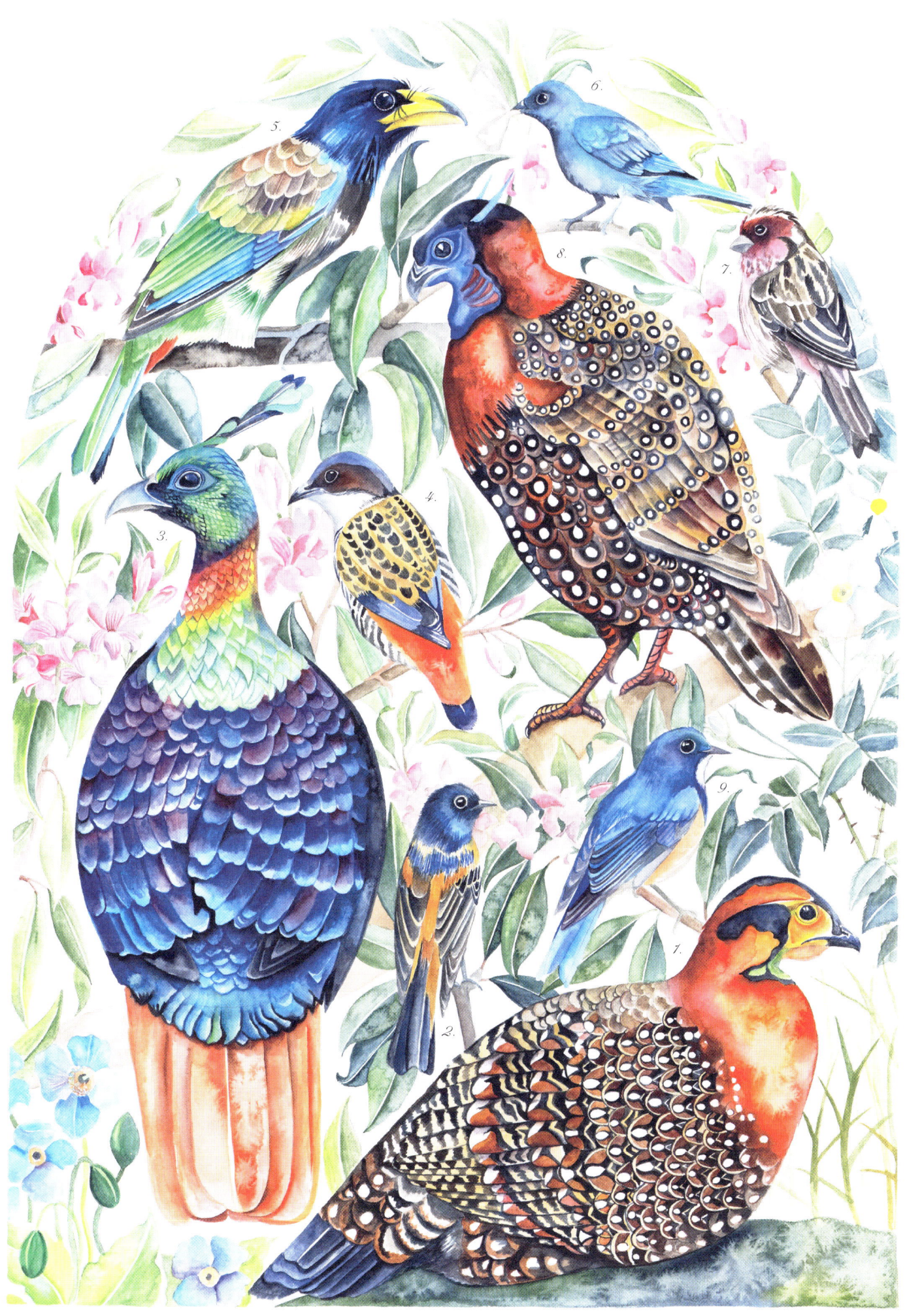

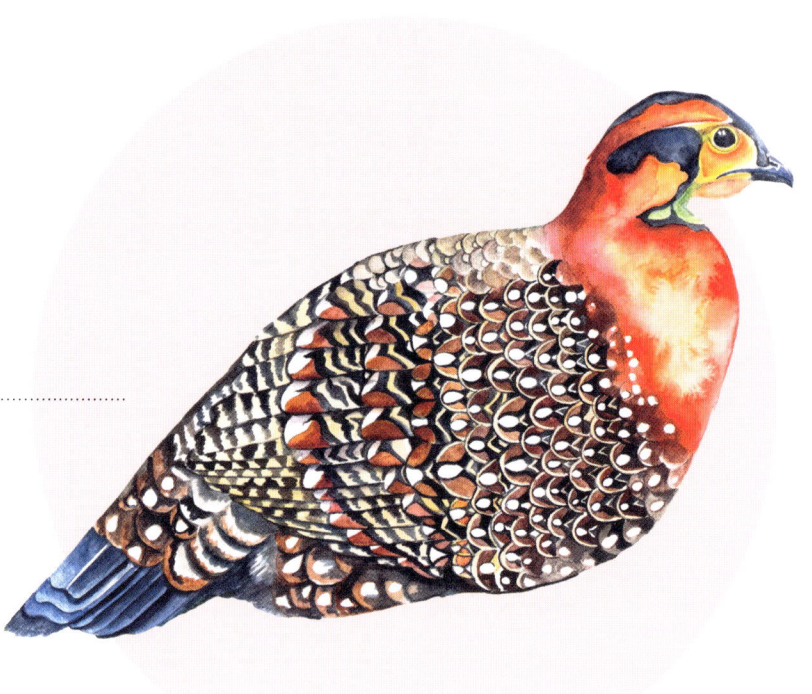

1.
BLYTH'S TRAGOPAN

Tragopan blythii
Order Galliformes

Like other tragopans, during mating season, the male has two blue "horns" on its head that become erect, and a colorful flap, called a lappet, hangs from its throat and is exposed to entice females.

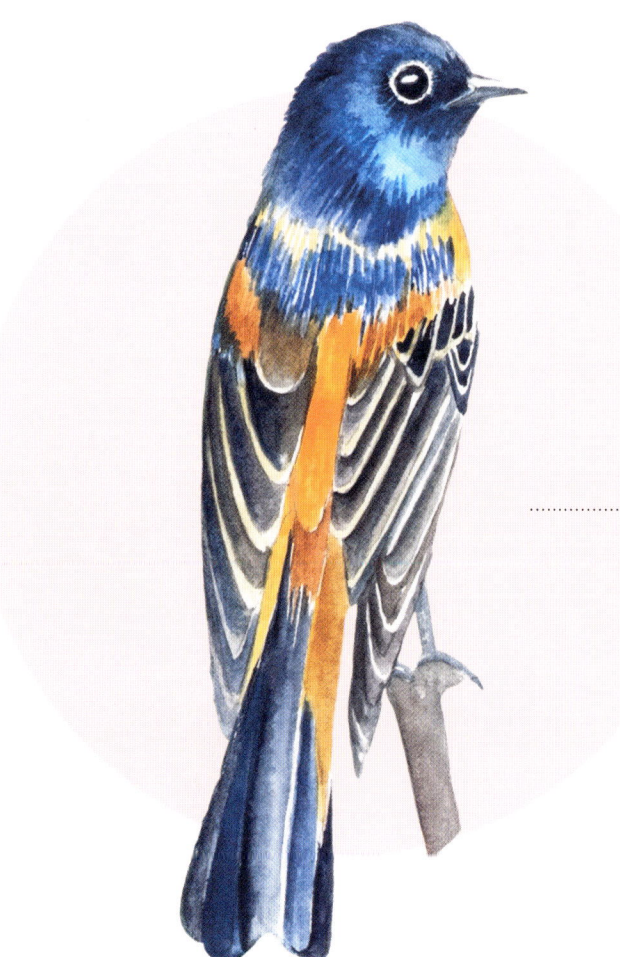

2.
BLUE-FRONTED REDSTART

Phoenicurus frontalis
Order Passeriformes

The male of this species has a cobalt-blue forehead and mantle and orange-red belly. These Old World flycatchers breed throughout the Himalayas and in China, where males are not quite as bright blue.

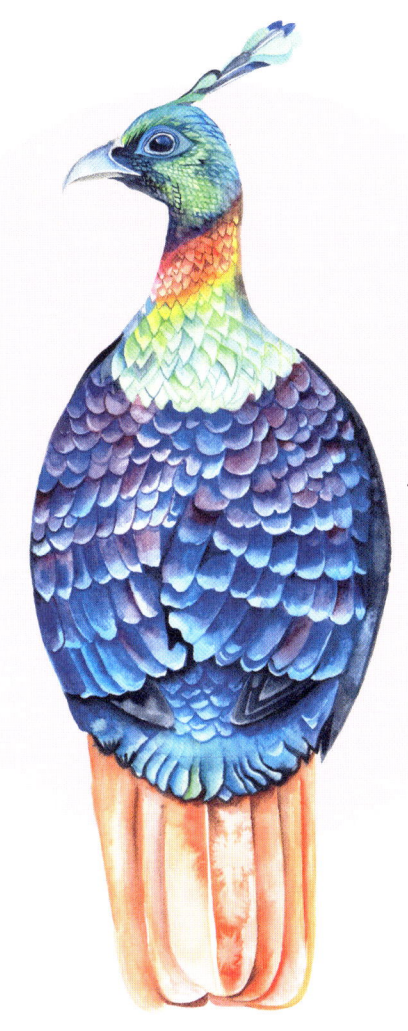

3.
HIMALAYAN MONAL

Lophophorus impejanus
Order Galliformes

After mating, the female scrapes a nest into the ground, where she lays three to five eggs. Down-covered chicks, which hatch with their eyes open, are completely independent after six months.

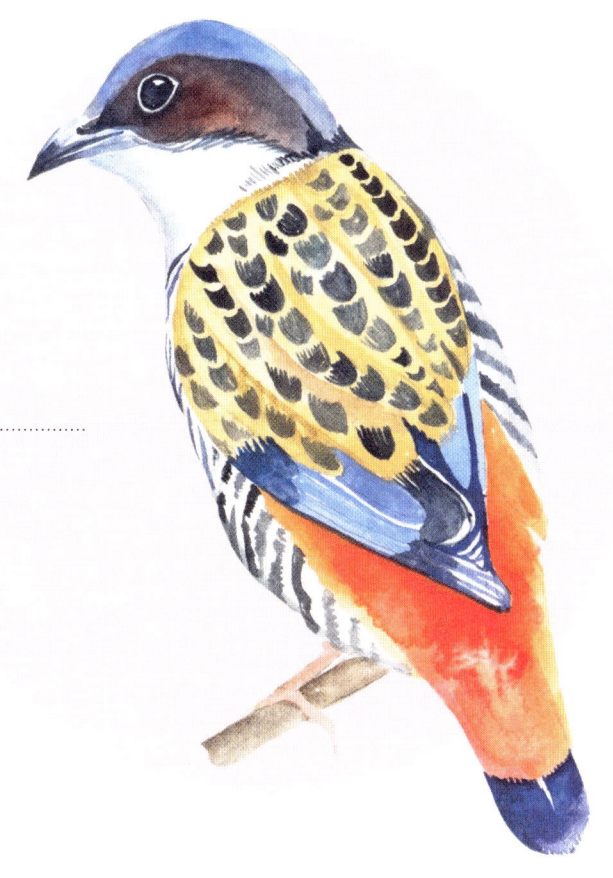

4.
HIMALAYAN CUTIA

Cutia nipalensis
Order Passeriformes

The female features beautiful zebra-like stripes of black-and-white feathers along her sides, and her tawny colored back is streaked with black spots. Males are similar, but with a solid rufous back.

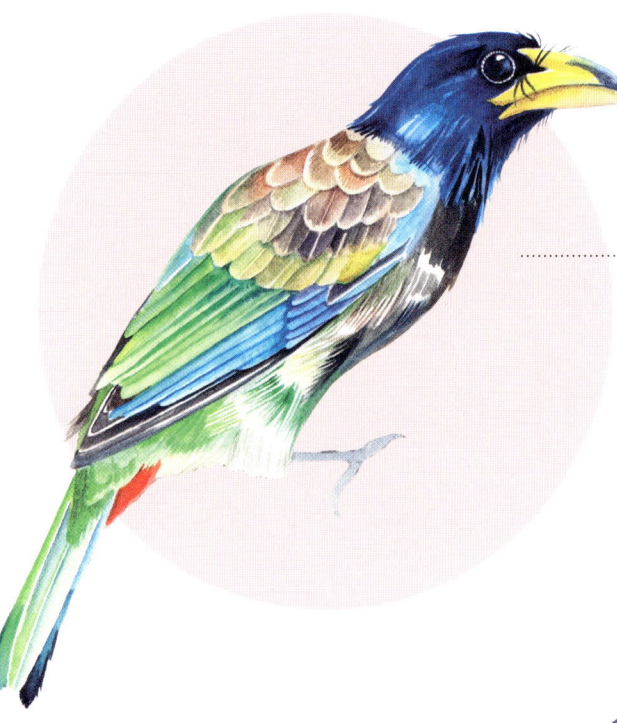

5.
GREAT BARBET

Psilopogon virens
Order Piciformes

This species is the largest of all barbets, with a body length of more than 14 inches. Its colors are also unique for a barbet.

6.
VERDITER FLYCATCHER

Eumyias thalassinus
Order Passeriformes

Unlike other flycatchers that prefer to perch closer to the ground, this iridescent blue species tends to forage above the canopy and perches on exposed treetop branches.

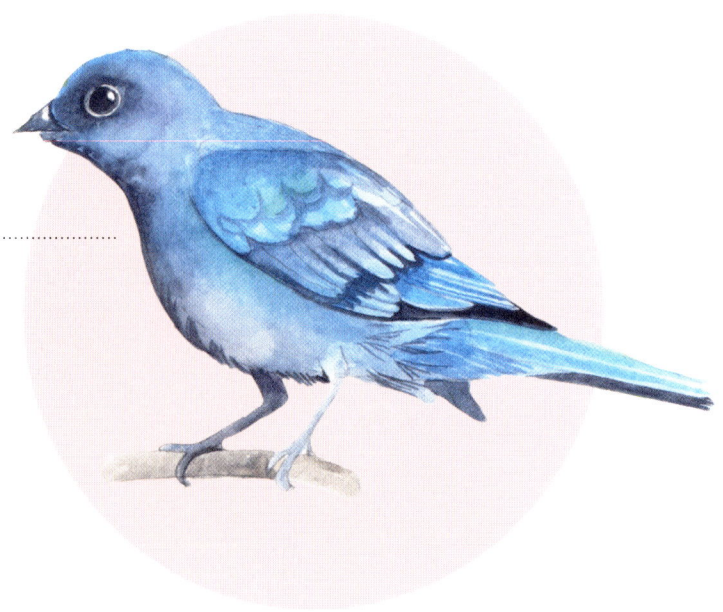

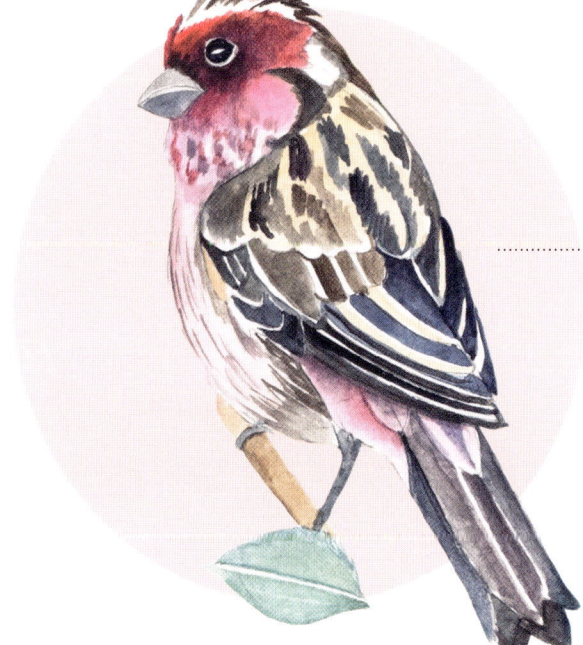

7.
HIMALAYAN WHITE-BROWED ROSEFINCH

Carpodacus thura
Order Passeriformes

Like other finches, this pink-and-brown-feathered alpine and subalpine species forages on the ground and in low bushes and feeds primarily on buds, shoots, and berries.

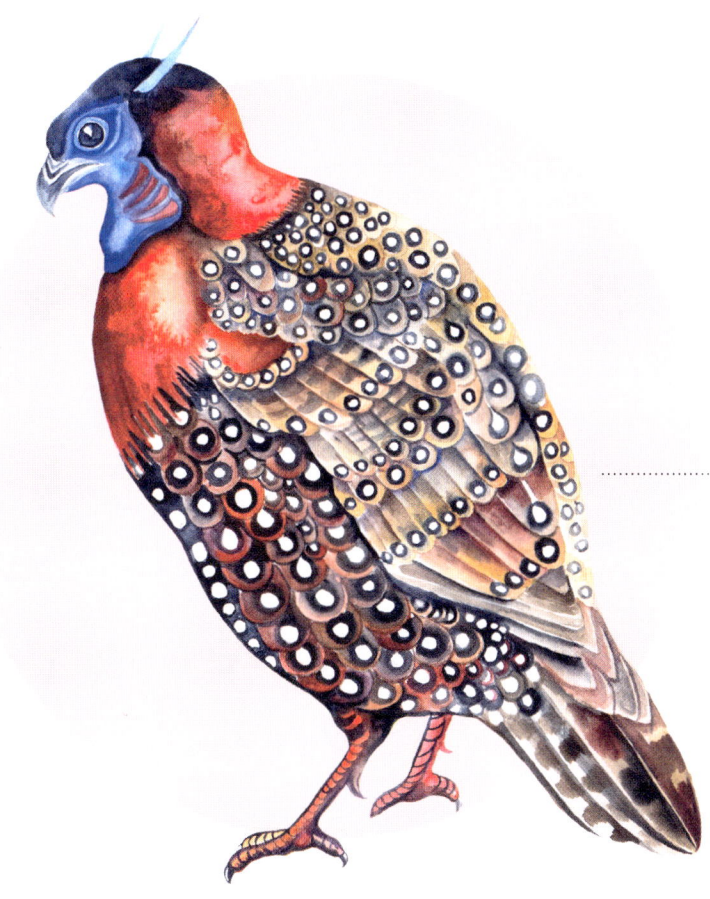

8.
WESTERN TRAGOPAN

Tragopan melanocephalus
Order Galliformes

This large, short-tailed pheasant is sometimes
called the "horned tragopan" because of
its distinctive bluish black occipital crest
of hornlike, elongated feathers.

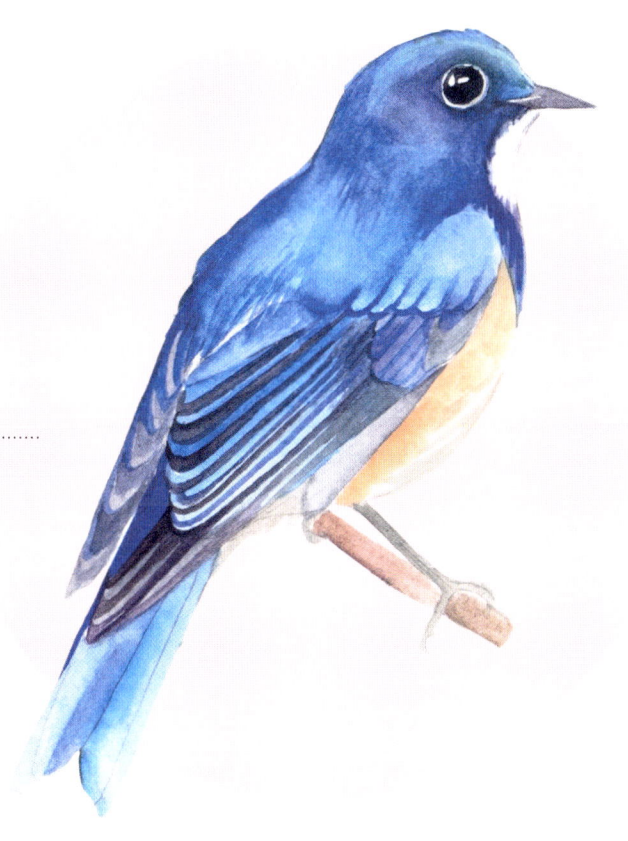

9.
HIMALAYAN BLUETAIL

Tarsiger rufilatus
Order Passeriformes

The male's sapphire-blue feathers and orange
flanks are easy to spot along forest tracks
in high-altitude forests during breeding
season or in edge habitats in winter.

107

BIRDS
from the
ISLAND of
BORNEO

The third largest island in the world, Borneo is covered in dense rainforest with a huge variation of flora, fauna, and bird species. This rainforest is believed to be about 140 million years old. To put this in perspective, the Amazon Rainforest has been in existence for a mere 55 million years!

The island's relative isolation has resulted in an ideal environment for many endemic bird species to thrive. Borneo has been (and continues to be) a crucial research site for various scientists, such as nineteenth-century British biologist Alfred Wallace, who extensively studied and illustrated the island's birds, plants, and animals. His research contributed to the theory of evolution by developing the concept of natural selection.

Search among the trees for the Rhinoceros Hornbill (4), easy to identify by its exceptionally large, conspicuously curved bill, which is topped by a prominent golden horn called a casque. The hollow and lightweight casque is made of keratin—the same protein found in human fingernails. Its calls echo into the casque's chamber, amplifying the sounds significantly. The Indigenous Iban people of Borneo hold this bird in extremely high regard, believing it to be an intermediary between humans and the spirit world. In the Iban cosmology, these birds are associated with warfare and are used in various religious ceremonies and practices.

1.
BORNEAN BRISTLEHEAD

2.
BORNEAN
SPIDERHUNTER

3.
RED-BEARDED
BEE-EATER

4.
RHINOCEROS HORNBILL

5.
STORK-BILLED
KINGFISHER

6.
BLUE-HEADED PITTA

7.
WHITEHEAD'S TROGON

8.
GOLDEN-NAPED
BARBET

9.
BORNEAN PEACOCK-
PHEASANT

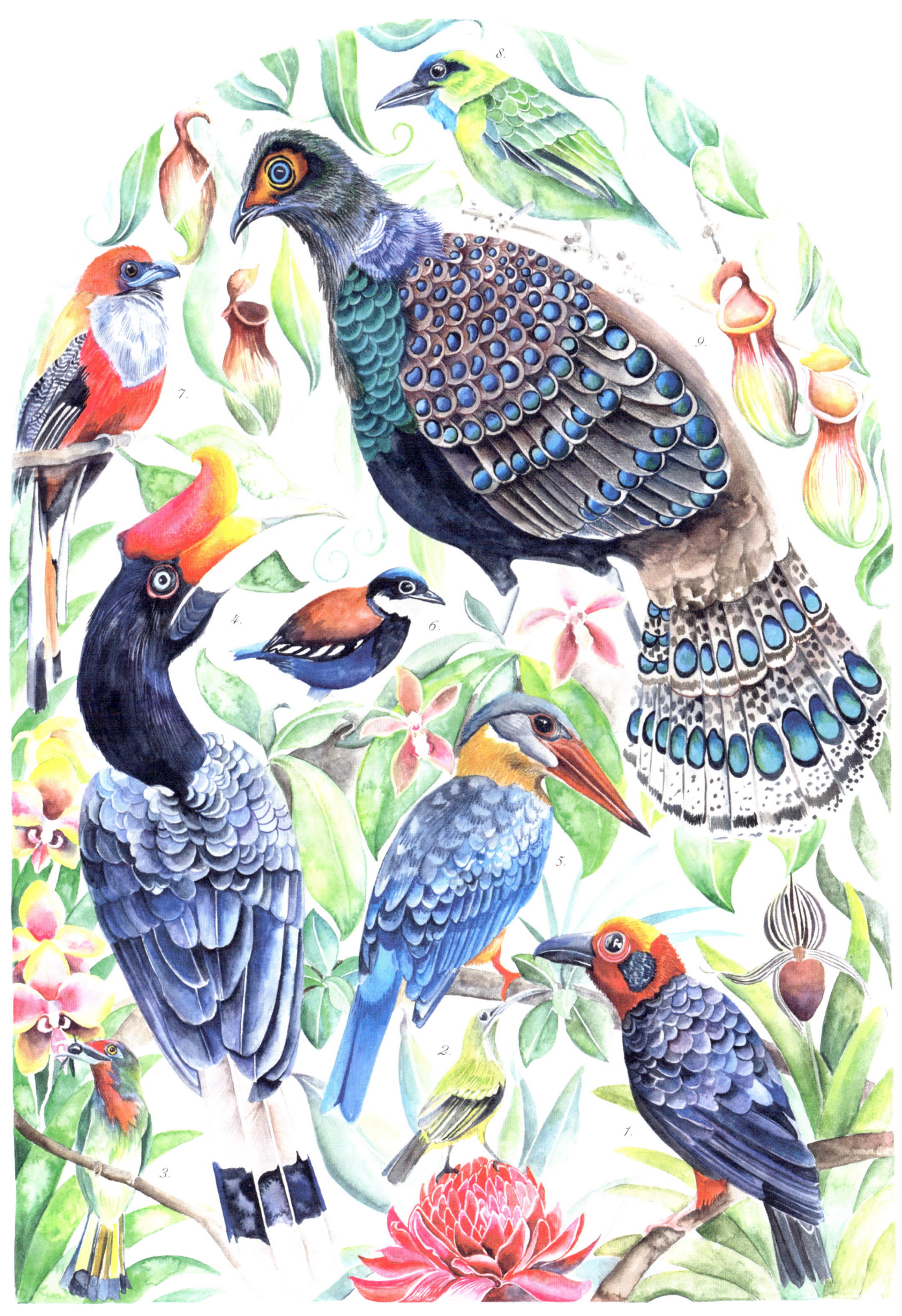

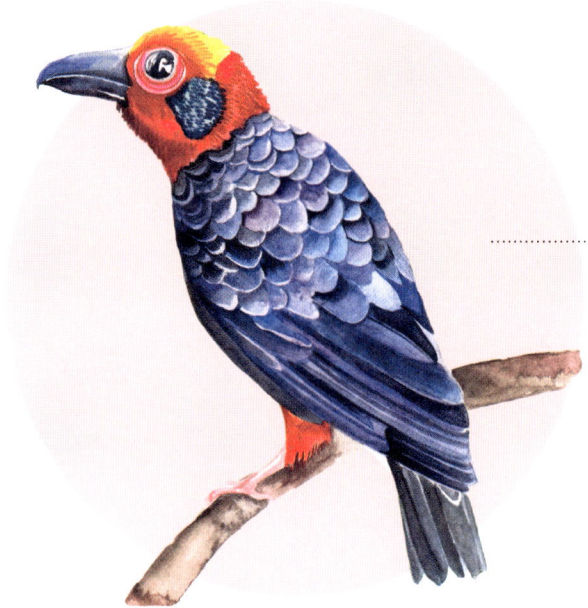

1.
BORNEAN BRISTLEHEAD

Pityriasis gymnocephala
Order Passeriformes

This bird is endemic to Borneo but sightings are rare.
They usually gather in a small flock and hop around
in low trees in peat swamps or mangrove forests.

2.
BORNEAN
SPIDERHUNTER

Arachnothera everetti
Order Passeriformes

Another species endemic to Borneo, the
spiderhunter tends to favor the nectar of luscious
blossoms of banana trees and wild ginger.

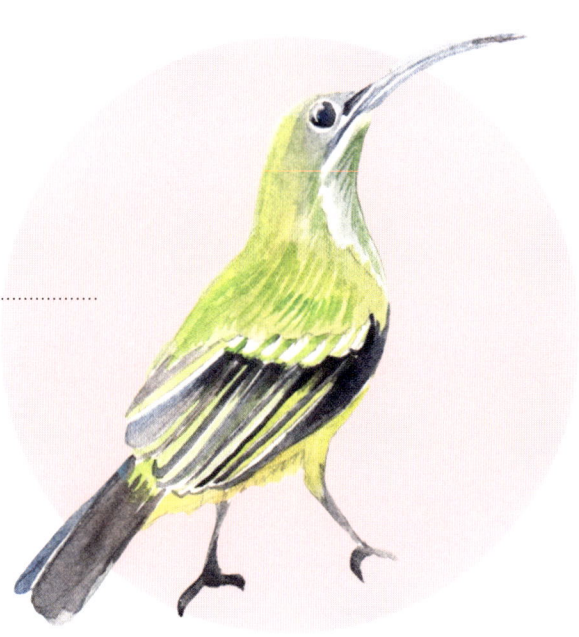

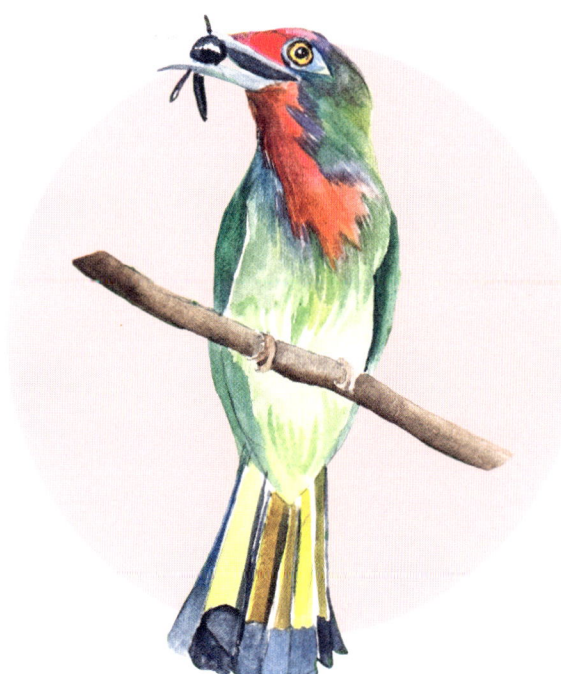

3.
RED-BEARDED BEE-EATER

Nyctyornis amictus
Order Coraciiformes

Although this species is brightly colored, it can
be extremely difficult to find. But its unique
descending, croaking call may be heard.

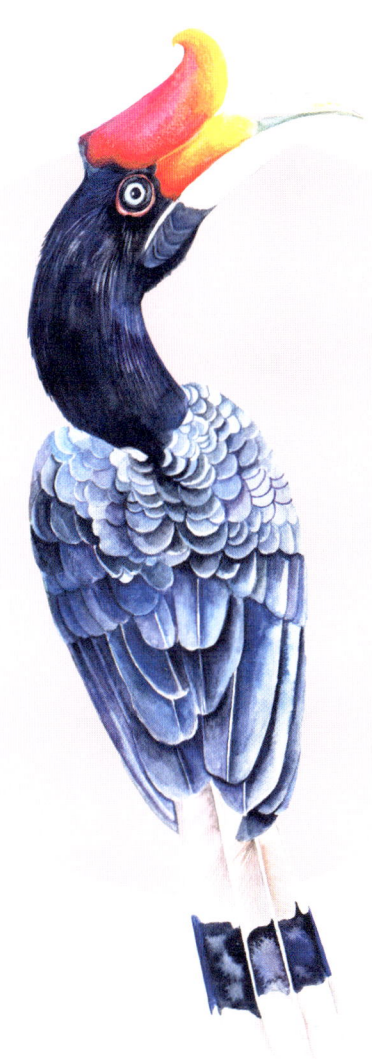

4.
RHINOCEROS HORNBILL

Buceros rhinoceros
Order Bucerotiformes

This majestic species with an enormous bill is one of the largest avian residents in all the rainforests of Southeast Asia. It can measure up to 41 inches long and can live for up to thirty-five years.

5.
STORK-BILLED KINGFISHER

Pelargopsis capensis
Order Coraciiformes

This enormous kingfisher is quite territorial and is known to chase away large raptors and other predators as it defends its nesting and hunting sites.

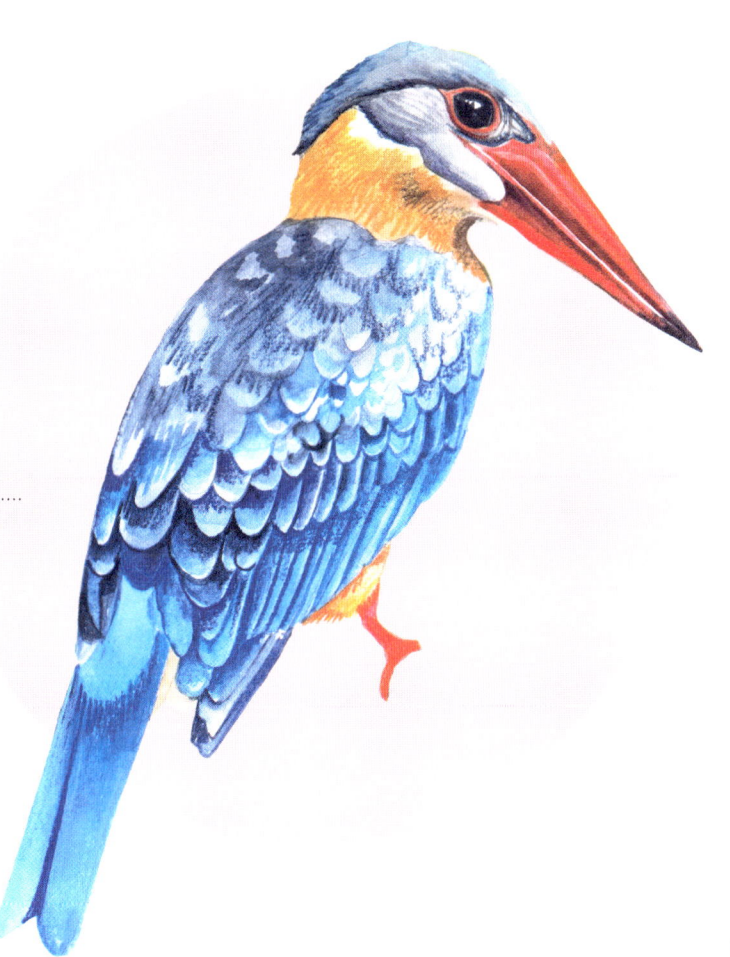

6.
BLUE-HEADED PITTA

Hydrornis baudii
Order Passeriformes

This jewel-toned species is found only in Borneo,
where it is threatened by habitat loss as the trees
in its forest home are being rapidly harvested.

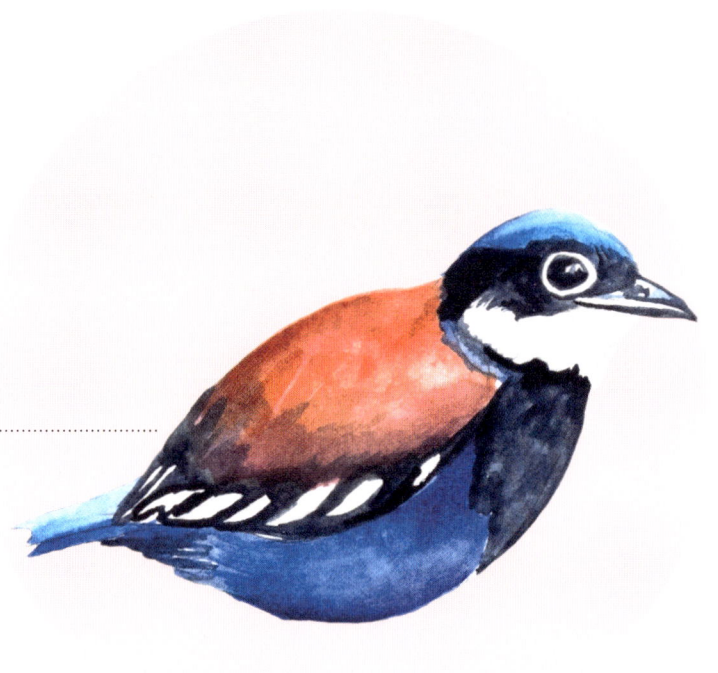

7.
WHITEHEAD'S
TROGON

Harpactes whiteheadi
Order Trogoniformes

Measuring upwards of 12 inches long, the largest
trogon species lives in the mountains of Borneo.
This brightly colored, shy bird tends to sit quietly
in the upper part of the dense canopy.

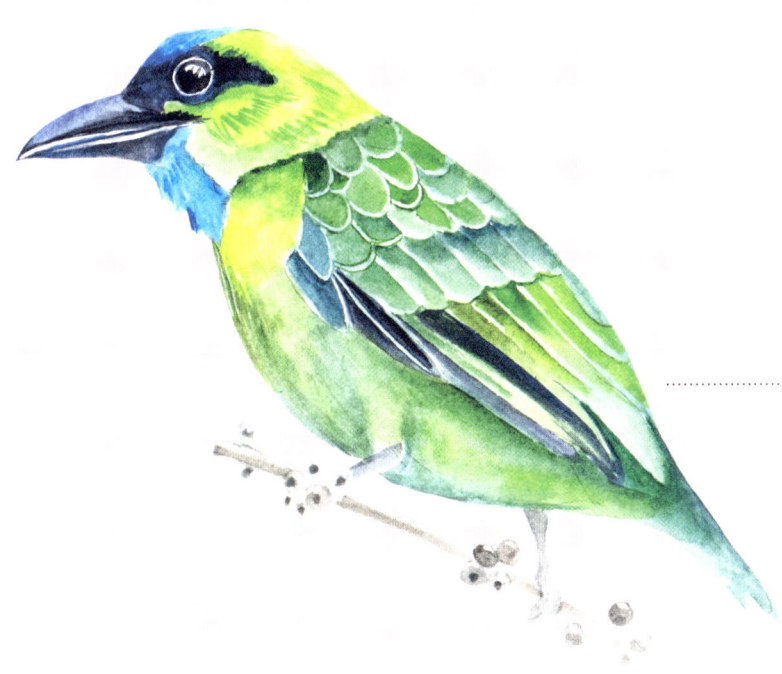

8.
GOLDEN-NAPED BARBET

Psilopogon pulcherrimus
Order Piciformes

This emerald-green fruit-eater is endemic to Borneo, where it forages in the midstory forest layer and nests in tree cavities.

9.
BORNEAN PEACOCK-PHEASANT

Polyplectron schleiermacheri
Order Galliformes

The rarest of all pheasants is also endemic to Borneo. Males of this technicolor species have an iridescent pale blue crest, bright orange facial skin patch, and luminous blue spots on the wings and tail.

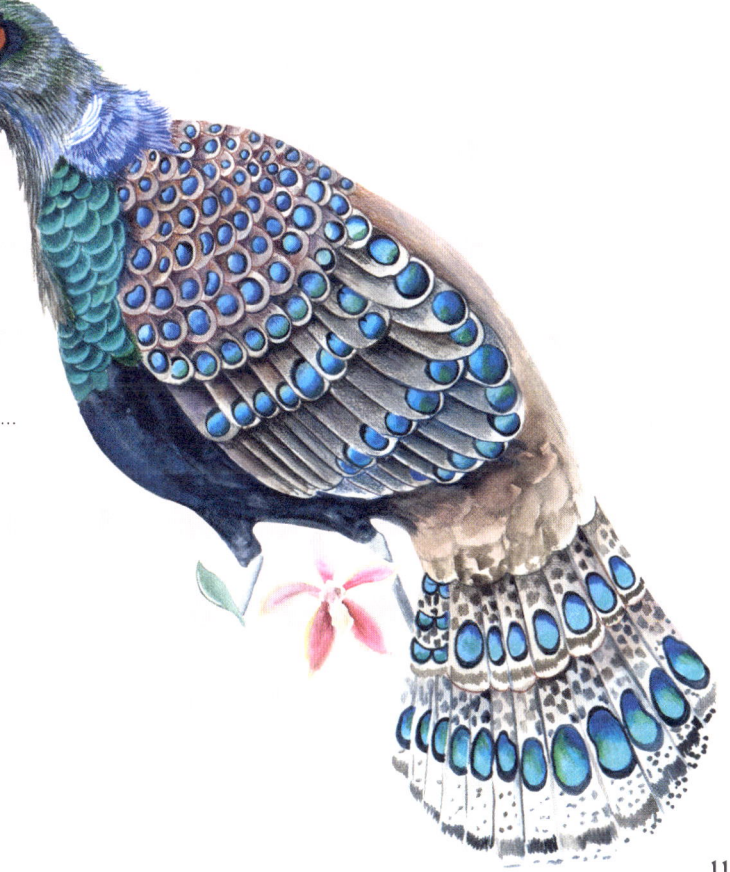

BIRDS
of the
INDONESIAN
ARCHIPELAGO

Indonesia is almost entirely tropical and home to many lush, abundant swathes of rainforest, towering volcanoes, glittering lakes, and sloping hills. Its rainforests are home to some of the richest biodiversity in the world, second only to Australia in the sum of its endemic species. Indonesia is a true birdwatcher's paradise, with a bounty of astonishing bird species.

One of the most remarkable birds is the bird-of-paradise, with forty-five species, two of which are featured here: Wilson's Bird-of-Paradise (1), which is endemic to Indonesia, and the Lesser Bird-of-Paradise (5), a New Guinea species that inhabits eastern Indonesia. In the sixteenth century, skins of the Greater Bird-of-Paradise were brought back to Spain from New Guinea by Ferdinand Magellan's crew members, with their wings and feet removed to showcase the vibrant plumage. The locals told the explorers that these birds drifted miraculously through the clouds without wings or feet, descending to the earth only in death, when they fell from the sky. This mythology persisted into the nineteenth century, when further exploration enabled naturalists to observe the birds in the wild, noting that they did, in fact, have wings and feet—but the name stuck.

1.
WILSON'S BIRD-OF-PARADISE

2.
JAVAN GREEN-MAGPIE

3.
SUMATRAN TROGON

4.
MALAYAN BANDED-PITTA

5.
LESSER BIRD-OF-PARADISE

6.
JAVAN KINGFISHER

7.
BLACK-AND-RED BROADBILL

8.
KNOBBED HORNBILL

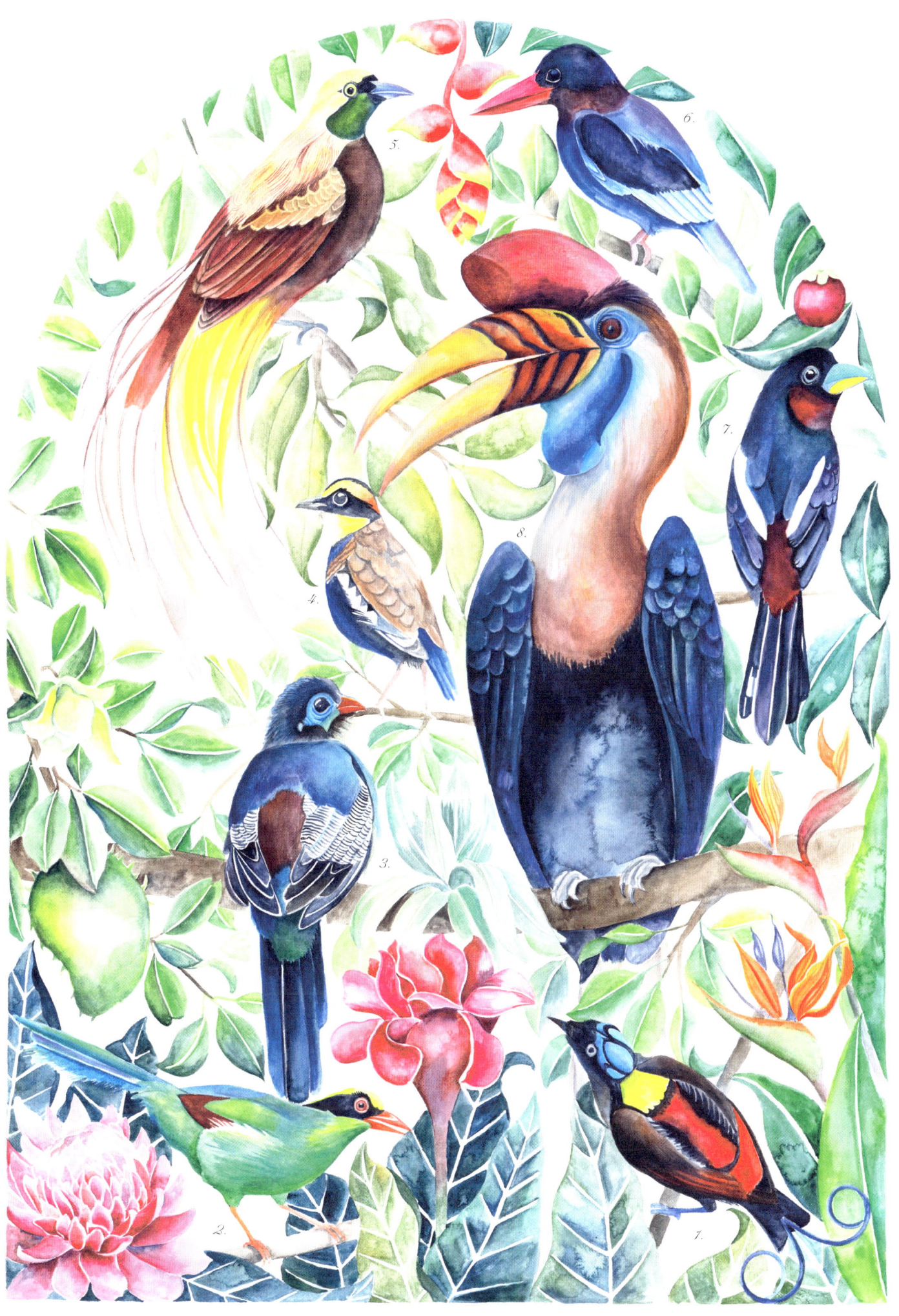

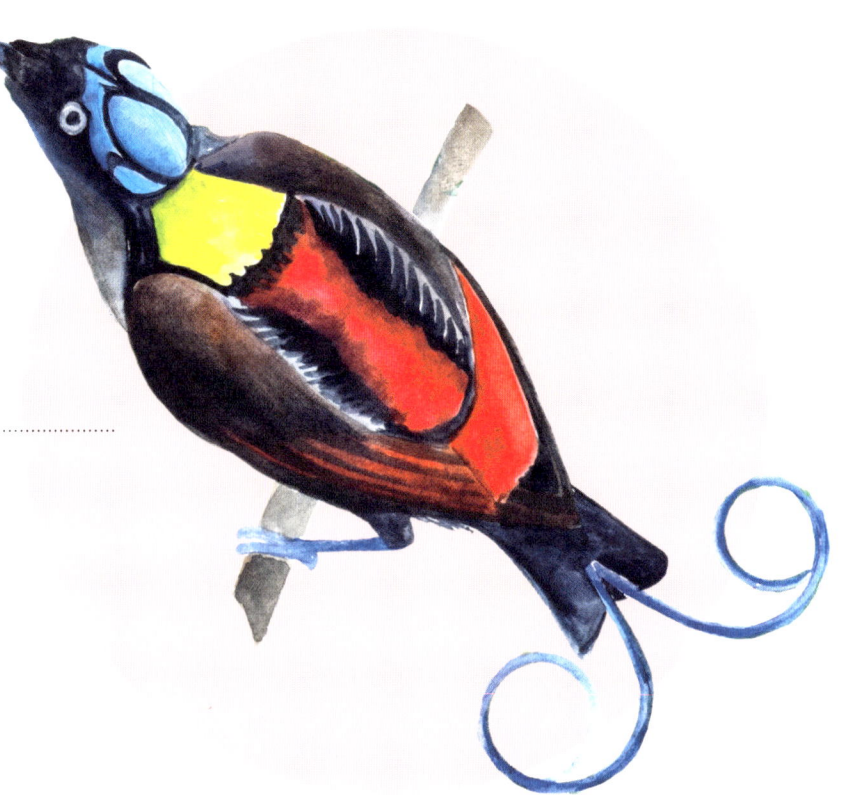

1.
WILSON'S
BIRD-OF-PARADISE

Diphyllodes respublica
Order Passeriformes

Males flash their bright feathers in an elaborate mating dance to impress females. They often clear debris from an area of the rainforest to set the stage, not wanting anything to distract from their own bright colors and movements.

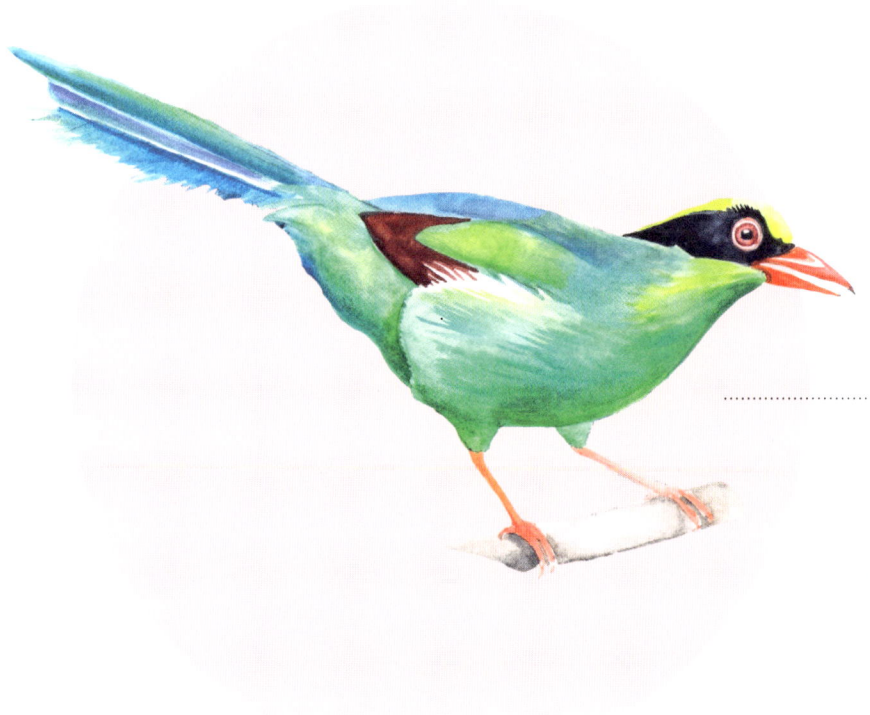

2.
JAVAN GREEN-MAGPIE

Cissa thalassina
Order Passeriformes

This vivid-green endemic Indonesian species is critically endangered and nearing extinction as a result of habitat loss and illegal capture for the wild animal trade.

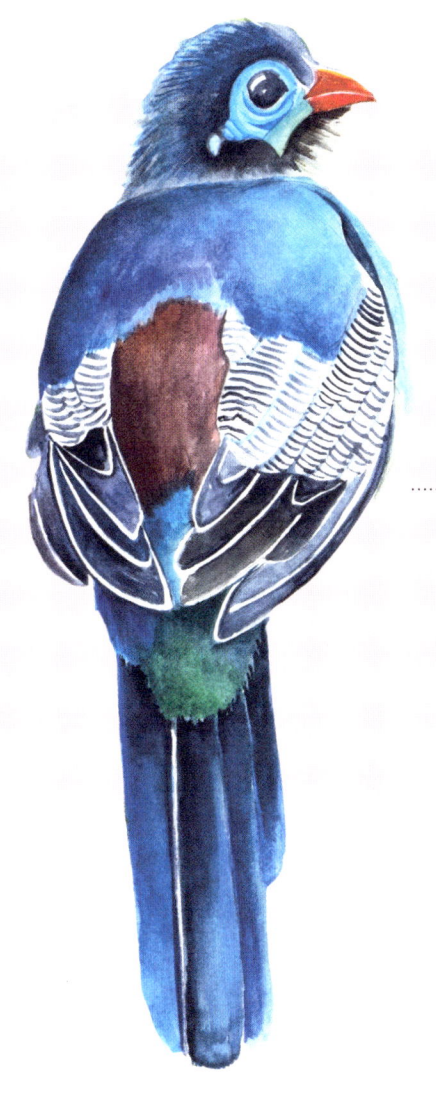

3.
SUMATRAN TROGON

Apalharpactes mackloti
Order Trogoniformes

This slow-moving, colorful bird lives deep
in the mountain forests of the island
of Sumatra, where it is endemic.

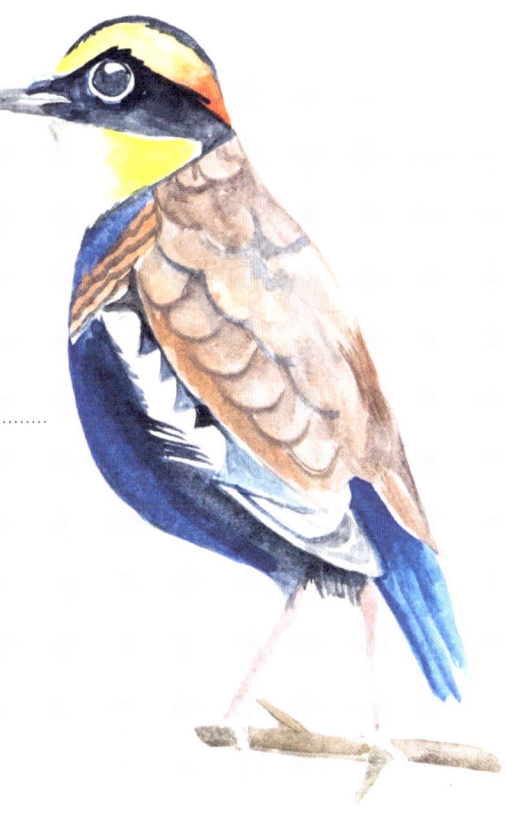

4.
MALAYAN
BANDED-PITTA

Hydrornis irena
Order Passeriformes

This species, the Bornean Banded-Pitta, and
the Javan Banded-Pitta were once considered
to be variants of the same species. In 2010,
scientists determined that they were three
separate species based on morphological,
ecological, and vocal differences.

5.
LESSER
BIRD-OF-PARADISE

Paradisaea minor
Order Passeriformes

Polyamorous males of this species perform courtship displays in areas called leks on the forest floor. They breed with numerous females. Each female usually lays two pinkish eggs in a nest high in a tree.

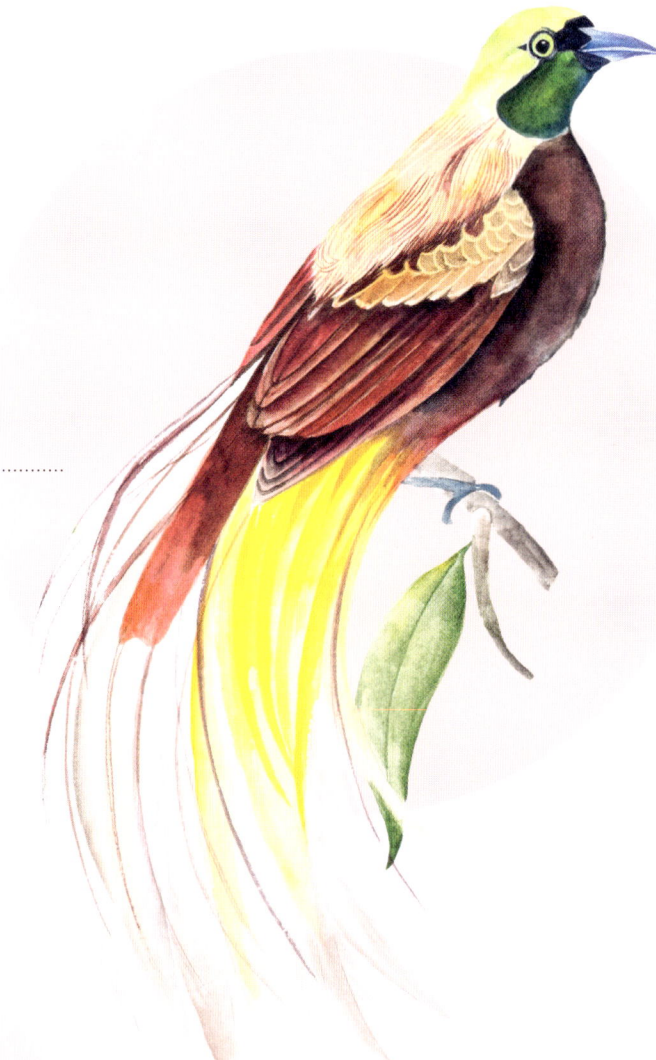

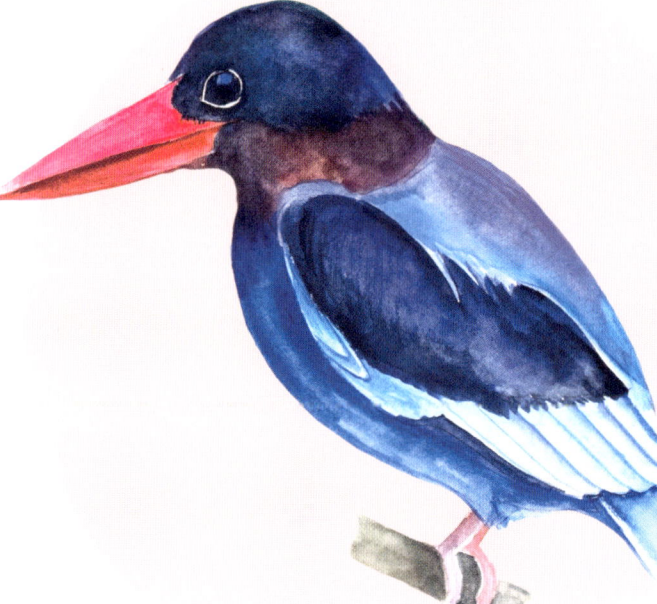

6.
JAVAN KINGFISHER

Halcyon cyanoventris
Order Coraciiformes

This species has a huge, lipstick-red bill and electric-blue wings. Its preferred prey includes crabs, frogs, and small lizards. It sometimes beats its prey against a branch to kill or soften it before devouring it.

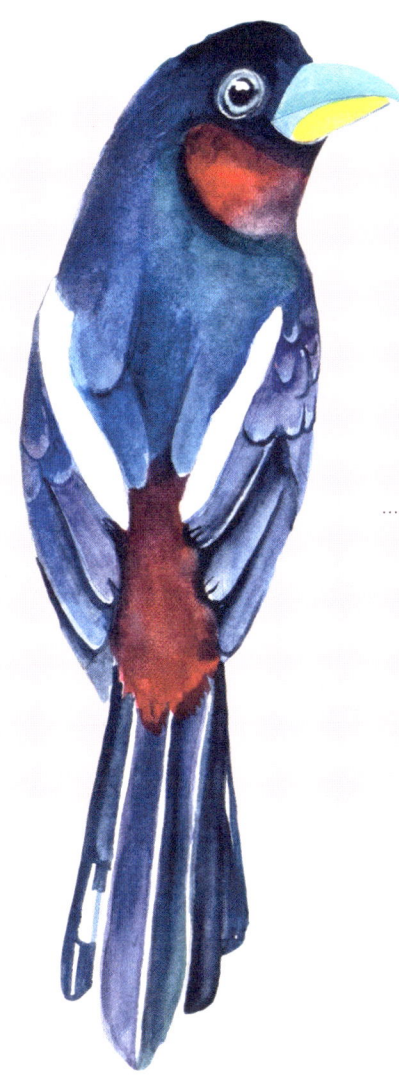

7.
BLACK-AND-RED BROADBILL

Cymbirhynchus macrorhynchos
Order Passeriformes

The bulbous turquoise bill of this bird makes it easy to identify. During breeding season, pairs build pendulous nests overhanging forested waterways.

8.
KNOBBED HORNBILL

Rhyticeros cassidix
Order Bucerotiformes

This endemic Indonesian bird lives in forests. Like the Northern Red-billed Hornbill, the female seals herself inside a tree cavity using her own droppings before laying her eggs. The male delivers food through a slit in the seal.

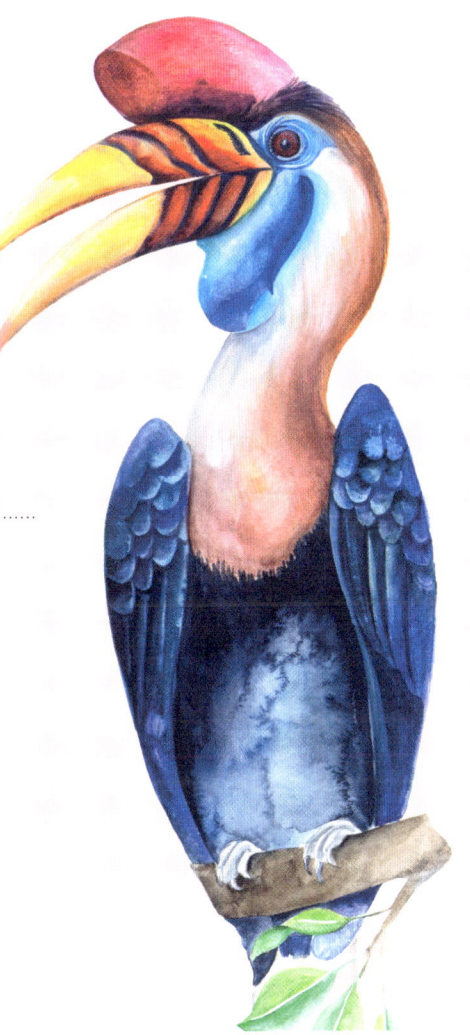

BIRDS of JAPAN

Japan, with a climate that ranges from subarctic to subtropical, is home to many endemic bird species and migratory visitors. The most iconic bird from Japan is undoubtably the Red-crowned Crane (2), an exceptionally large and elegant bird with brilliant white plumage. Featured in Japanese art for centuries, the crane holds tremendous cultural significance in several other Asian countries including Korea, China, and Mongolia. In Japan, it is said that the sacred crane can live for more than 1000 years.

In one popular Japanese folktale, a poor young man saves an injured crane that falls from the sky. When the man returns home that evening, he finds a beautiful young maiden who says she is to become his wife. All she asks is that he build her a room of her own where she can weave, but he must promise never to enter. From her secret room, she produces exquisitely beautiful woven cloth, which the man sells. As time passes, the man notices that his wife is looking increasingly unwell. One day, overcome with curiosity and temptation, the man breaks his promise and enters the secret room. There he finds not his wife, but a white crane weaving the cloth, plucking and using its own feathers to make the delicate fabric. When he realizes this is the crane he once saved, the spell is broken and the bird flies away, leaving the sad young man heartbroken.

Another bird with a prominent place in Japanese mythology is the Green Pheasant (4), which in traditional folklore is considered a messenger between Heaven and Earth. It is the national bird of Japan and was once depicted on Japanese currency.

1.
JAPANESE QUAIL

2.
RED-CROWNED CRANE

3.
JAPANESE WOODPECKER

4.
GREEN PHEASANT

5.
JAPANESE BUSH WARBLER

6.
JAPANESE ROBIN

7.
RED-FLANKED BLUETAIL

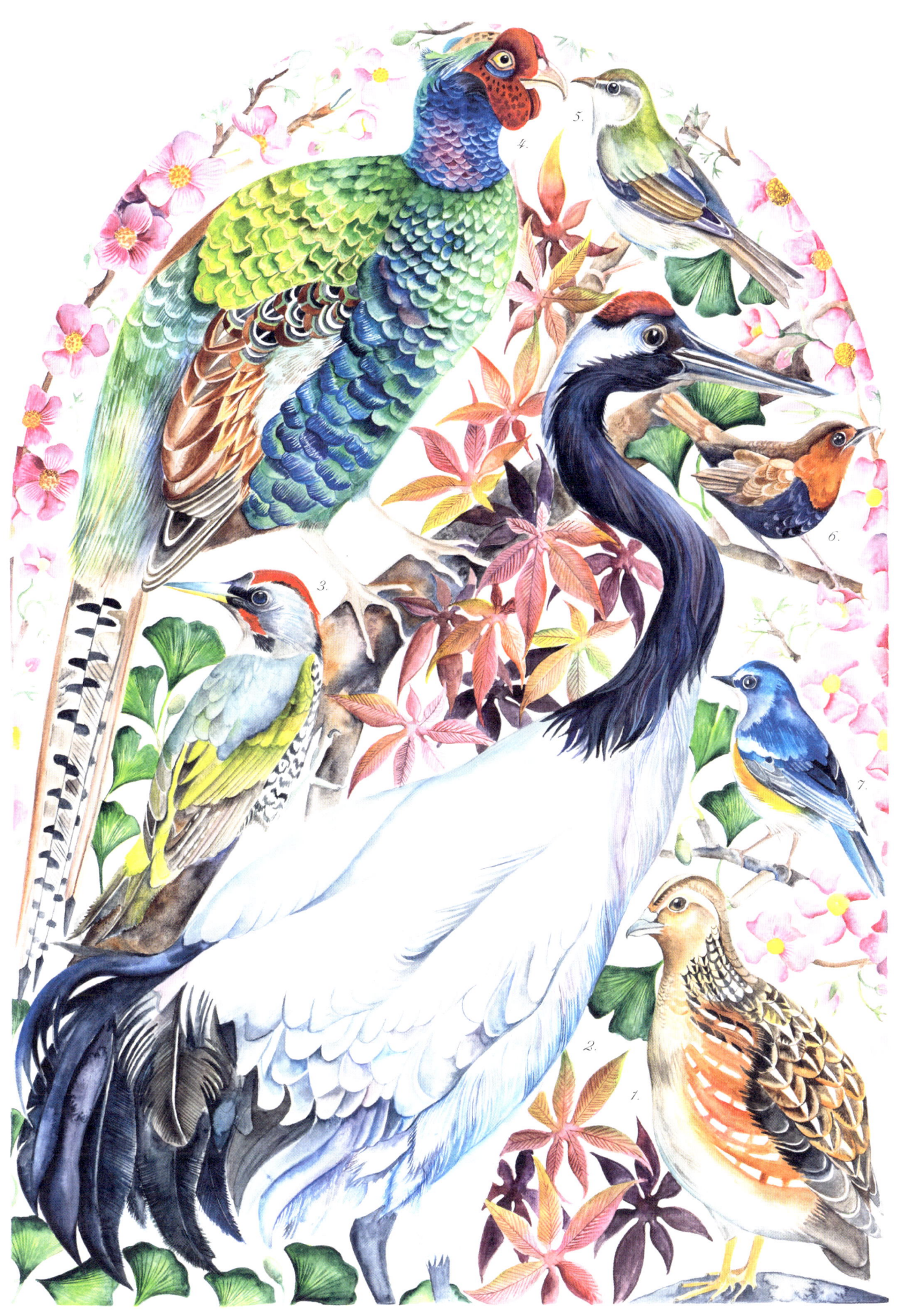

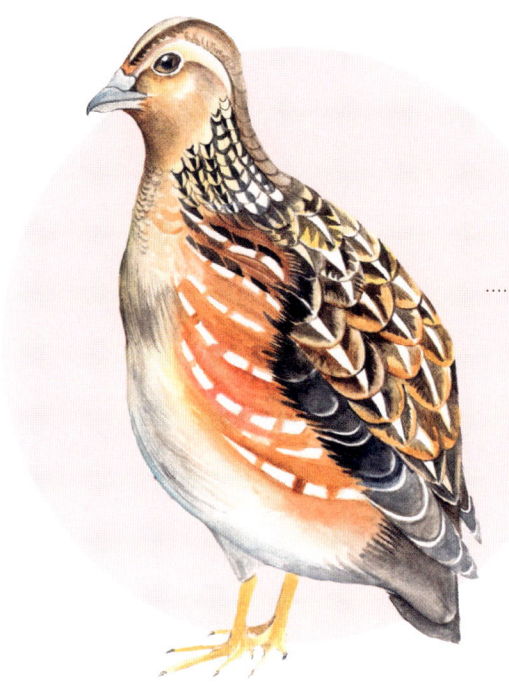

1.
JAPANESE QUAIL

Coturnix japonica
Order Galliformes

These shy meadow birds have been domesticated in Japan for egg and meat production since the eleventh century.

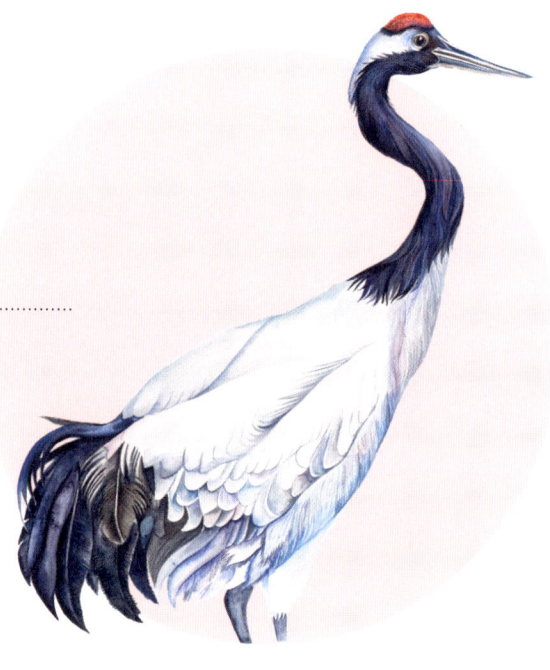

2.
RED-CROWNED CRANE

Grus japonensis
Order Gruiformes

The scarlet crown of this wetland bird becomes brighter and more distinctive during mating season.

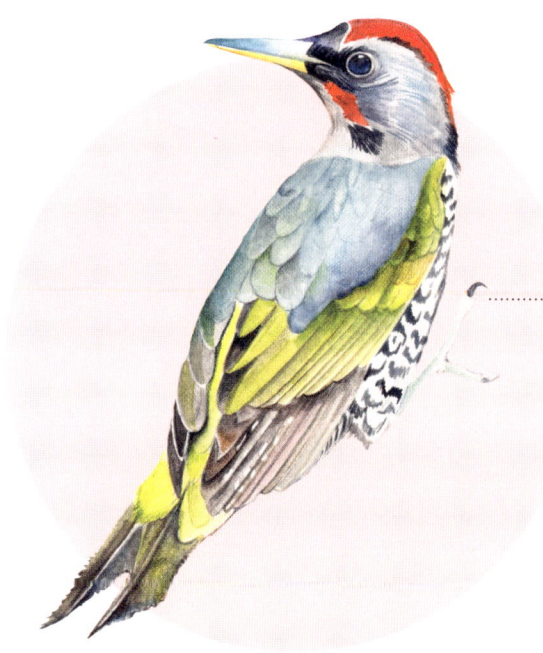

3.
JAPANESE WOODPECKER

Picus awokera
Order Piciformes

Although similar and closely related to the European Green Woodpecker, this species is a Japanese endemic.

4.
GREEN PHEASANT

Phasianus versicolor
Order Galliformes

In Japan, some people claim that Green Pheasants scream when they sense that an earthquake is imminent.

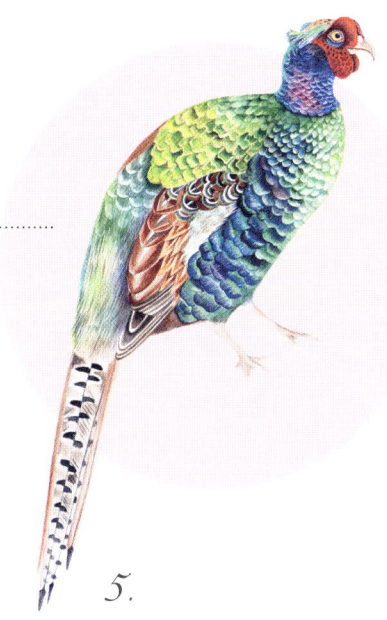

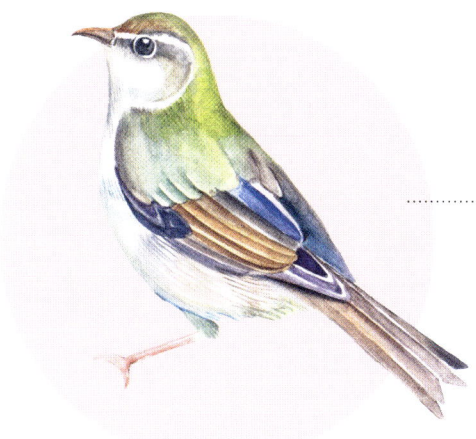

5.
JAPANESE BUSH WARBLER

Horornis diphone
Order Passeriformes

This plain brown warbler has a uniquely beautiful song, and because it is secretive and difficult to locate, it is more often heard than seen.

6.
JAPANESE ROBIN

Larvivora akahige
Order Passeriformes

Like the Japanese Bush Warbler, this little songbird is elusive and far more often identified by its distinctive call than by sightings.

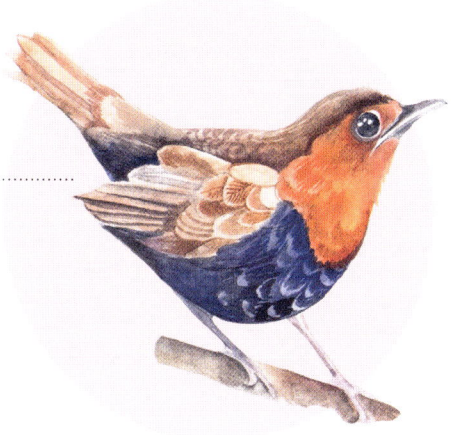

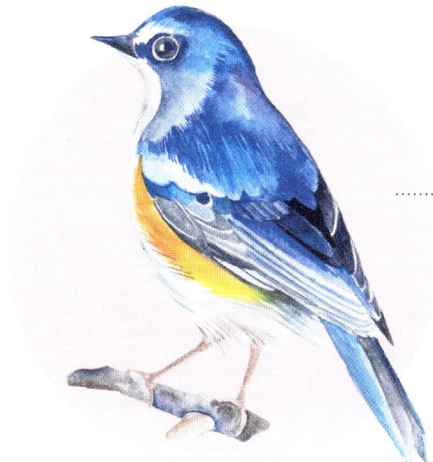

7.
RED-FLANKED BLUETAIL

Tarsiger cyanurus
Order Passeriformes

While feeding on insects and berries, this hyperactive little blue flycatcher moves quickly, flying from its perch to the forest floor while frequently flicking its tail.

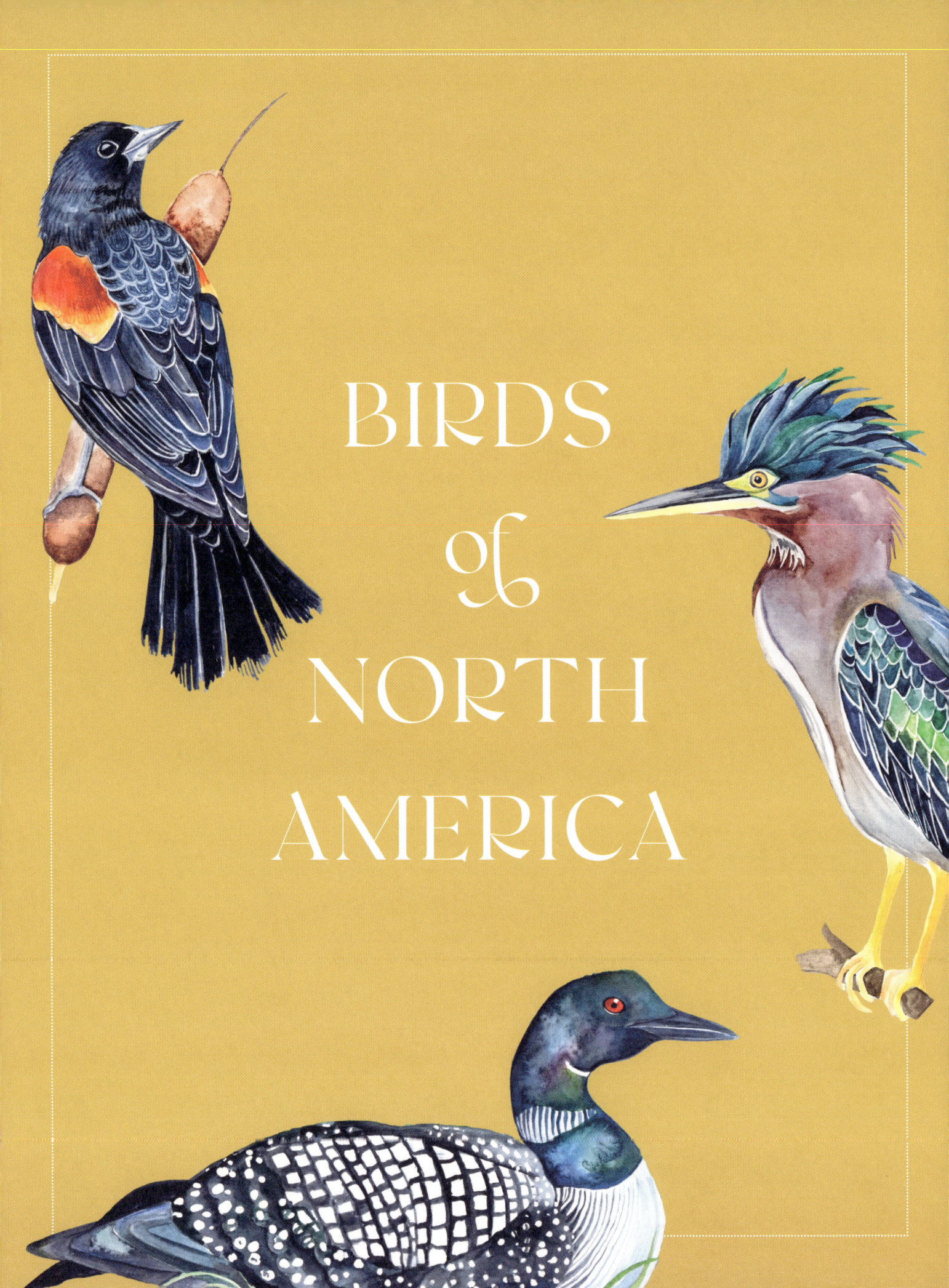

BIRDS
of
NORTH
AMERICA

It may be surprising to learn that flamboyant, colorful birds are not limited to the jungles and tropical rainforests of distant lands but can also be found throughout North America. In addition to familiar favorites like the American Robin, the Northern Cardinal, and the House Sparrow, birdwatchers may meet more exotic-looking birds like the Rose-breasted Grosbeak or the Common Loon. In the sinuous wetlands of the Southern United States, where Spanish moss hangs over the surface of alligator-infested waters, you may spot a Green Heron or a Purple Gallinule.

Scan the shoreline of the Florida coast for a Roseate Spoonbill—a gregarious pink bird whose unusual bill is used to detect crustaceans hiding in the mud and sand. Farther north, in the dewy forests of New England, you may be lucky enough to see the remarkable patterning of the Northern Flicker or the brilliantly colored and tempestuous Blue Jay. Head west and brave the sweltering heat of the desert, and you may see a roadrunner zipping across the landscape.

BIRDS
of the
CANADIAN
LAKES

In the northeastern wilds of Canada, amid ancient and towering pines, miles of rugged wilderness, gushing waterfalls, and crystalline lakes, are many beguiling and beautiful birds. With more freshwater lake area than any other country in the world, Canada is host to a variety of waterbirds, such as the graceful and otherworldly Common Loon (1). The ethereal call of a loon rising hauntingly from the mist above the still waters of a quiet lake is not a sound you'll soon forget. Loons require the clear, clean water of northern lakes for their nesting sites, and they depend on open expanses of water for the running starts they use to power their flight. The loon is a symbol of the wild, the cold North, and the undisturbed peaceful natural wonders that still exist.

While sleeping out under the clear, star-washed skies, you might be roused by the low hoot of a Barred Owl (4) perched atop a snarled pine. Owls have long been considered portents of danger and sorcery, and some Native American mythology refers to them as shapeshifters or reincarnated humans in the form of birds.

You may also meet the tufted, ruddy headed Common Merganser (6). An Ojibwe creation tale tells of the Earth being created by a waterfowl that, after a great flood, plunges to the bottom of the waters to emerge with a bit of earth within its beak, from which land is created. Sources diverge on which species of bird successfully brings forth the earth—in some telling, it is a merganser, while in others, it is a loon.

1.
COMMON LOON

2.
BLACK-THROATED
BLUE WARBLER

3.
RUFFED GROUSE

4.
BARRED OWL

5.
PILEATED
WOODPECKER

6.
COMMON MERGANSER

7.
NORTHERN PARULA

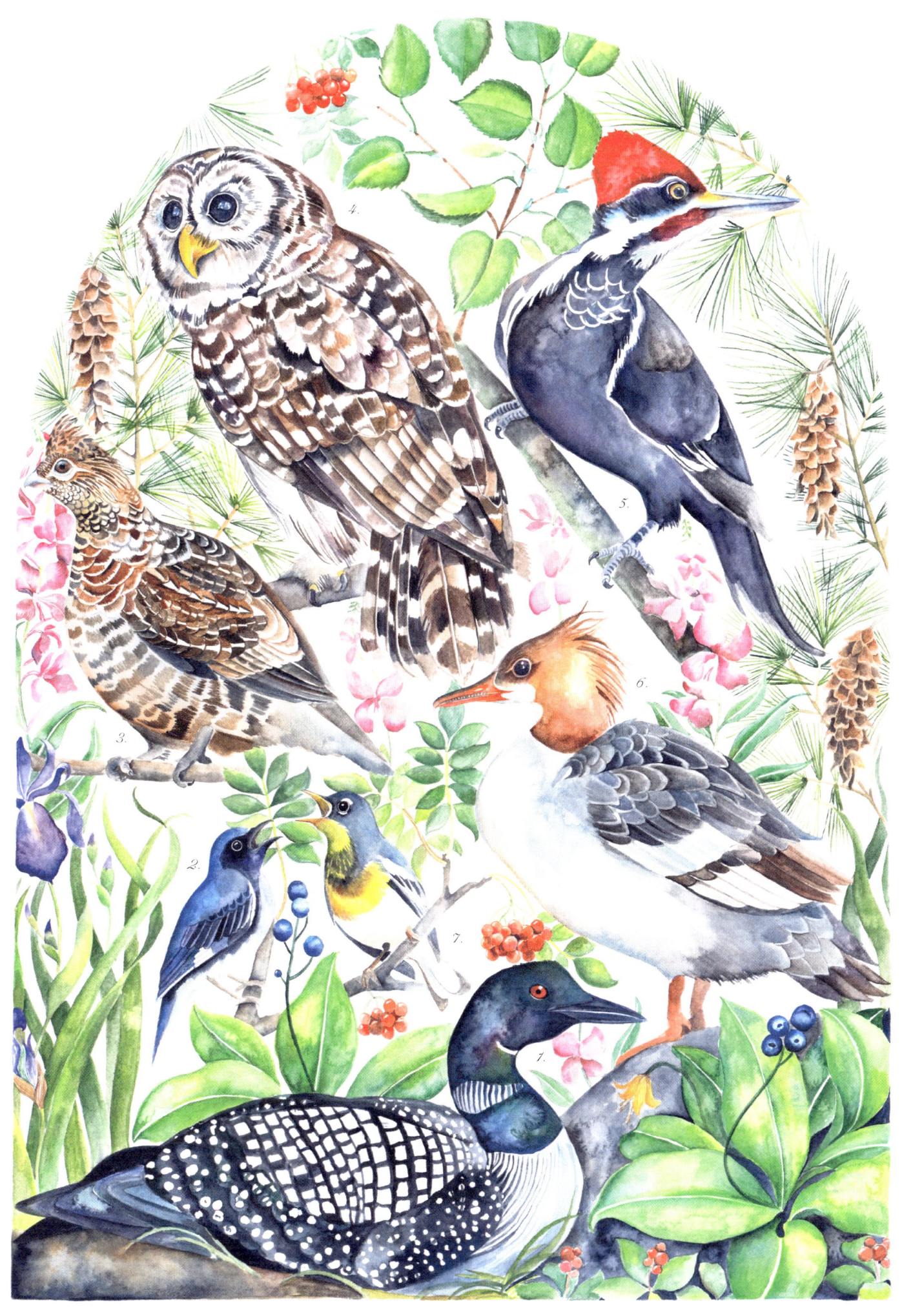

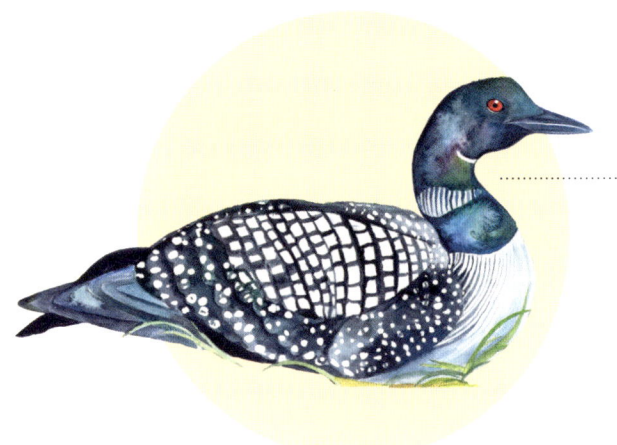

1.
COMMON LOON

Gavia immer
Order Gaviiformes

A loon's legs are positioned far back on its body, which, in addition to powerful webbed toes, makes it an adept swimmer but awkward on land. Loons go ashore only to mate and incubate eggs.

2.
BLACK-THROATED BLUE WARBLER

Setophaga caerulescens
Order Passeriformes

The common name for this species comes from the male's lovely blue plumage, but females are a plain olive-brown. The male and female look so different that they were once believed to be two different species.

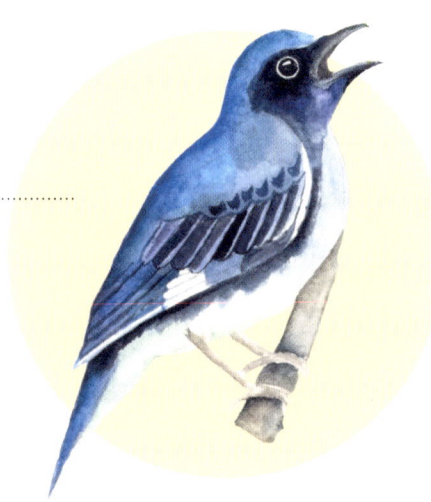

3.
RUFFED GROUSE

Bonasa umbellus
Order Galliformes

In winter, Ruffed Grouse grow comb-like projections off their toes, which are believed to act like snowshoes to help them walk across snowy terrain.

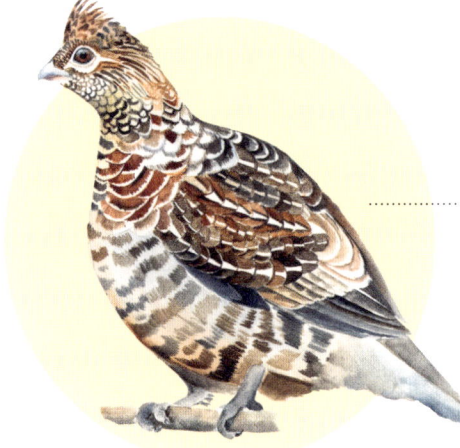

4.
BARRED OWL

Strix varia
Order Strigiformes

One of the most serious predatory threats to the Barred Owl is the larger and more aggressive Great Horned Owl.

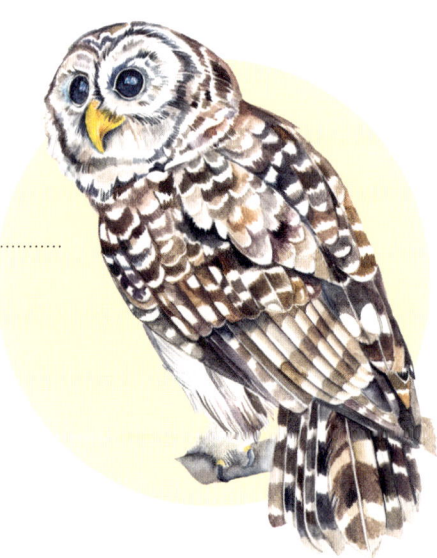

5.
PILEATED WOODPECKER

Dryocopus pileatus
Order Piciformes

This largest species of woodpecker in North America is one of the inspirations for the iconic cartoon character Woody Woodpecker.

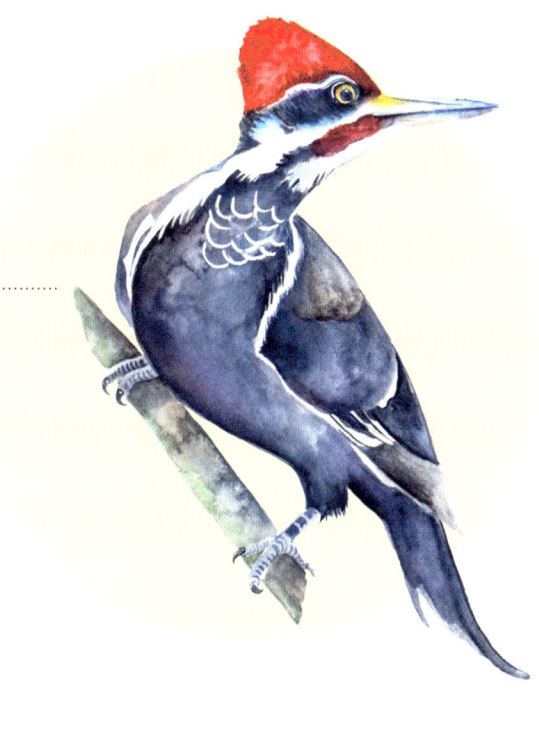

6.
COMMON MERGANSER

Mergus merganser
Order Anseriformes

This species' genus name comes from a Latin word that roughly translates to "plunging goose"—fitting for a bird that often dives underwater after prey.

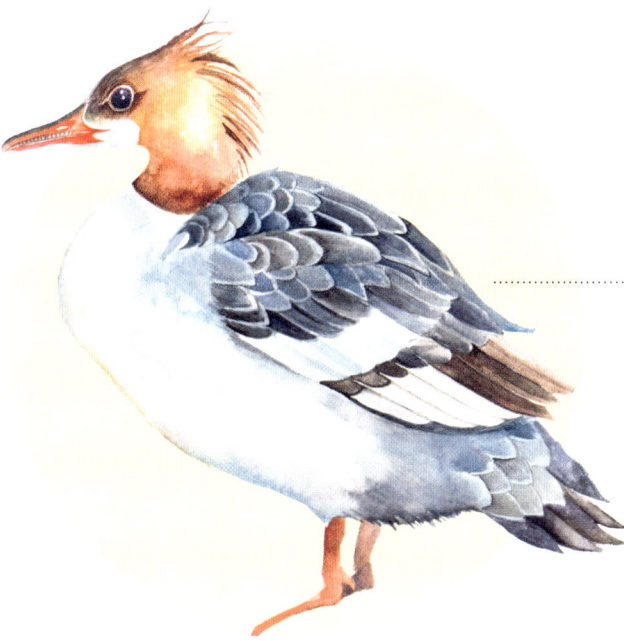

7.
NORTHERN PARULA

Setophaga americana
Order Passeriformes

This bird's breeding range is interestingly divided, extending from Florida to the Canadian boreal forest, but it skips parts of Iowa, Wisconsin, Michigan, and some Northeast states.

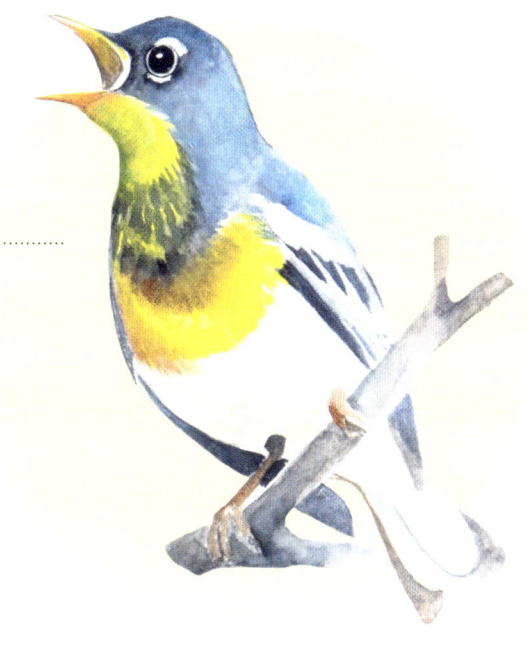

BIRDS
of
FLORIDA
WETLANDS

In the Southern United States, where graceful herons inhabit wetlands along with vicious crocodiles and deadly snakes, many remarkable birds make their home. Here you may see the only spoonbill in North America, the Roseate Spoonbill (3), a gregarious pink bird whose unusual bill and vibrant color make it look like something out of an imaginary landscape. Search among the water lilies and water hyacinths of a Southern wetland, and you may spot a Purple Gallinule (7), a remarkable flier that can migrate far beyond its normal range, from Switzerland to South Africa. Look for the shy and increasingly scarce Mangrove Cuckoo (9) as it wanders through dense vegetation in spring. This species' croaking call is said to predict a rainstorm, earning it the name "rain crow."

The distinctive Belted Kingfisher (8) is a remarkable diver, often seen perching above the water, scouting for fish. Kingfishers are found in many parts of the world, where they have inspired many stories and symbols. In the story of Noah's ark, for example, Noah sent out a dove in search of dry land and then released a kingfisher. When the kingfisher was freed from the ark, it was so filled with joy in the sunlight that it did not return. Some say that kingfishers sit looking into the water of wetlands, lakes, and streams as they await the ark's return.

1.	*3.*	*5.*	*7.*	*9.*
GRAY KINGBIRD	ROSEATE SPOONBILL	GREEN HERON	PURPLE GALLINULE	MANGROVE CUCKOO
2.	*4.*	*6.*	*8.*	
LEAST BITTERN	BLACK-CROWNED NIGHT HERON	PALM WARBLER	BELTED KINGFISHER	

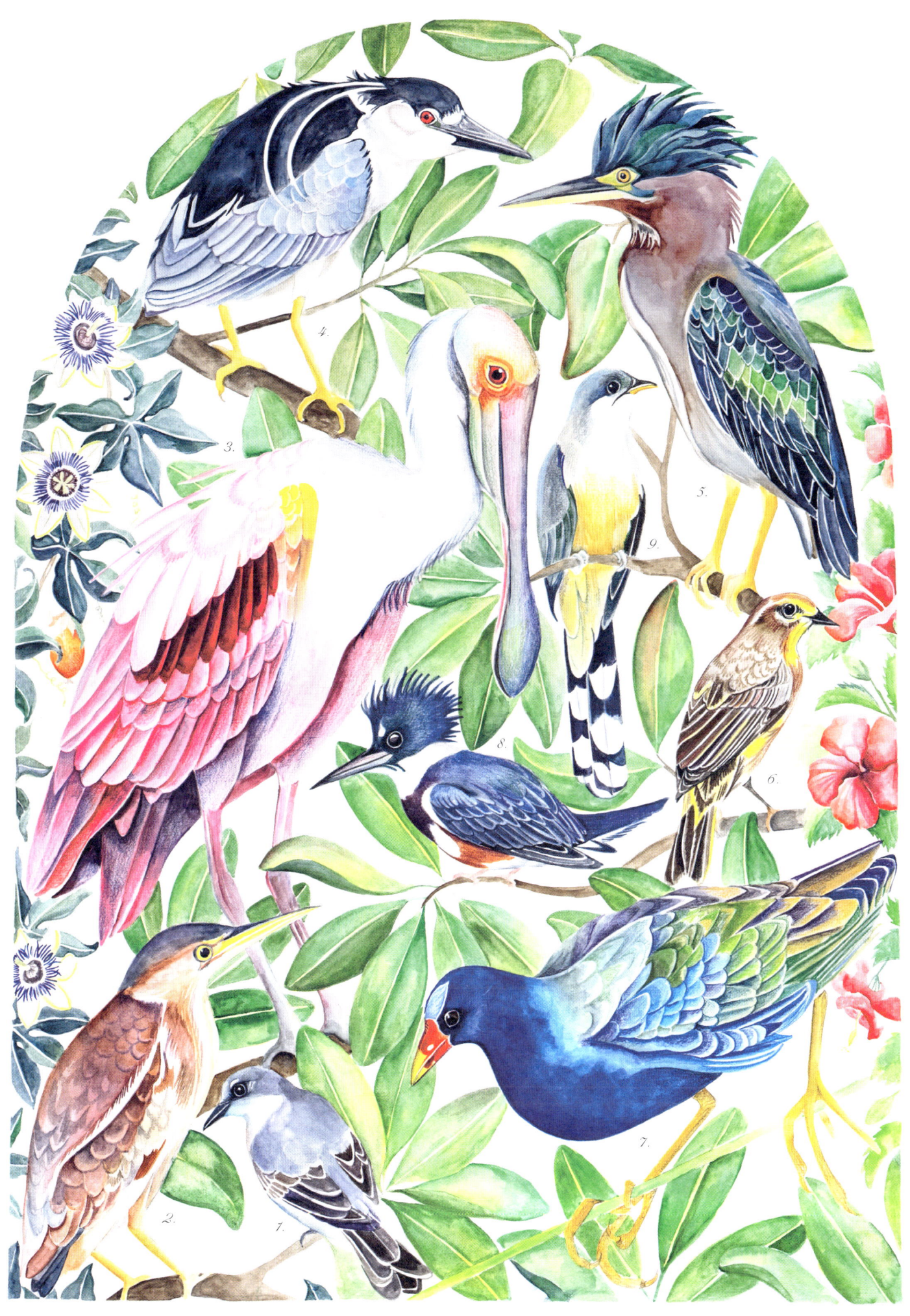

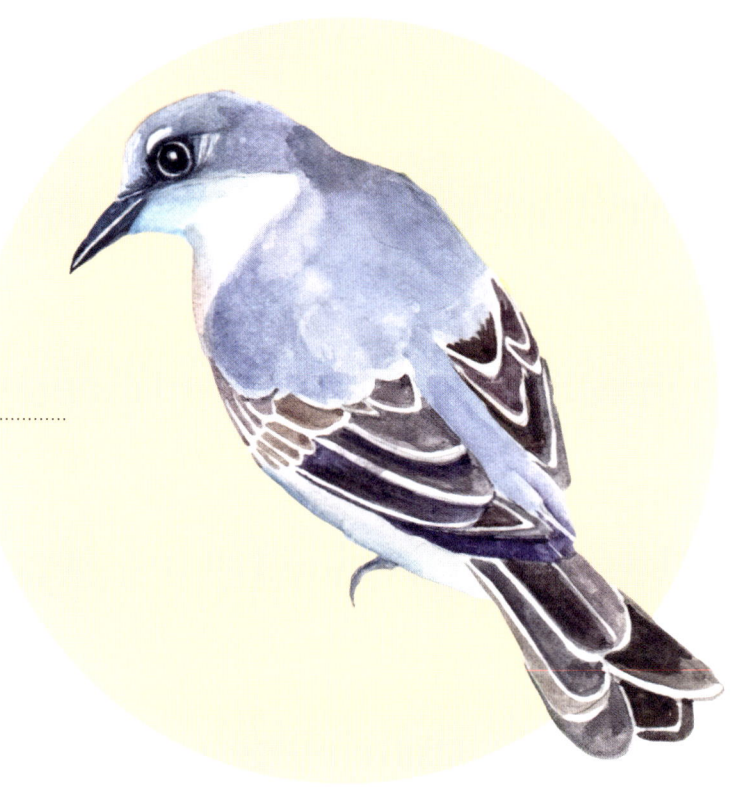

1.
GRAY KINGBIRD

Tyrannus dominicensis
Order Passeriformes

The male and female of this extremely
territorial coastal bird will aggressively
defend their nest against potential predators,
attacking dogs, pigs, and even humans.

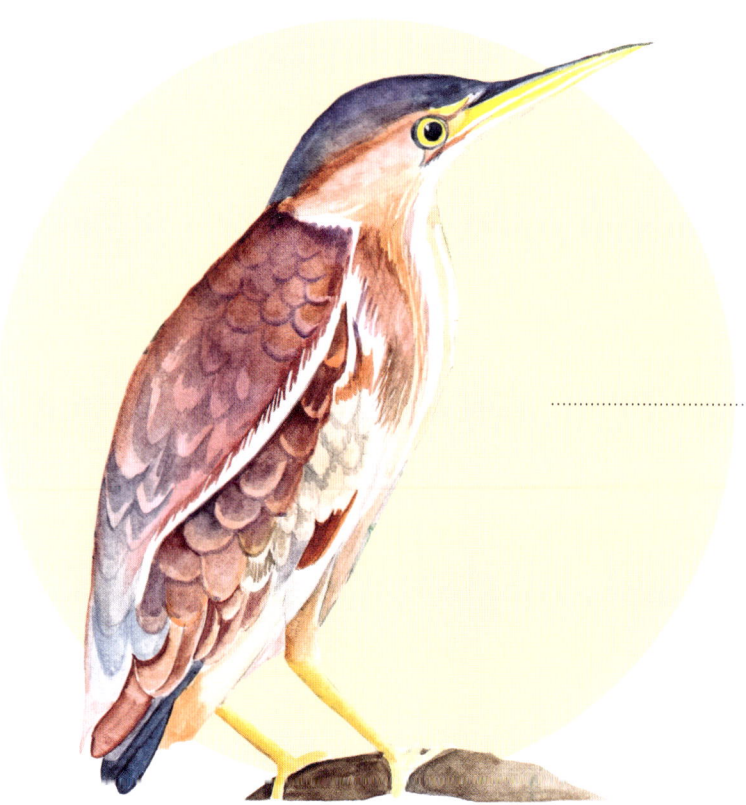

2.
LEAST BITTERN

Ixobrychus exilis
Order Pelecaniformes

When these marsh birds are startled or feel
threatened, they stay completely still and point
their beaks toward the sky to blend in. They'll
sometimes sway a bit, so that they resemble
marsh vegetation moving in the wind.

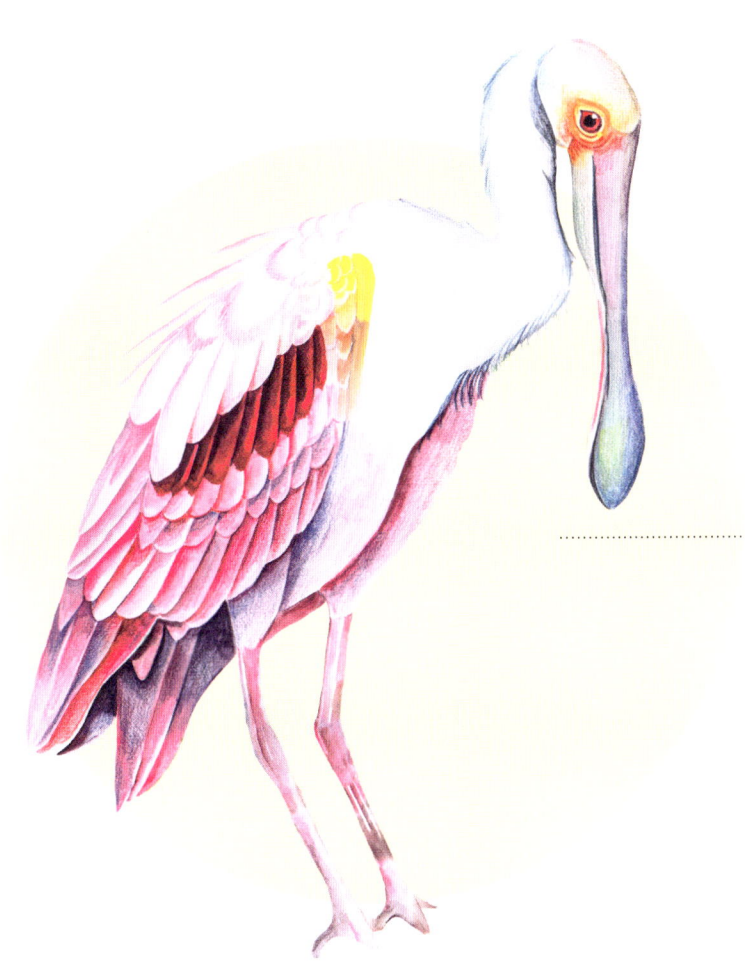

3.
ROSEATE SPOONBILL

Platalea ajaja
Order Pelecaniformes

Like flamingos, Roseate Spoonbills' feather colors
are derived from their diet. Crustaceans and other
aquatic invertebrates contain pigments called
carotenoids that can turn their feathers pink.

4.
BLACK-CROWNED
NIGHT HERON

Nycticorax nycticorax
Order Pelecaniformes

This species, the most widespread heron in
the world, will brood any chick that is placed
in its nest. Heron parents apparently don't
distinguish between their own nestlings
and offspring from other avian parents.

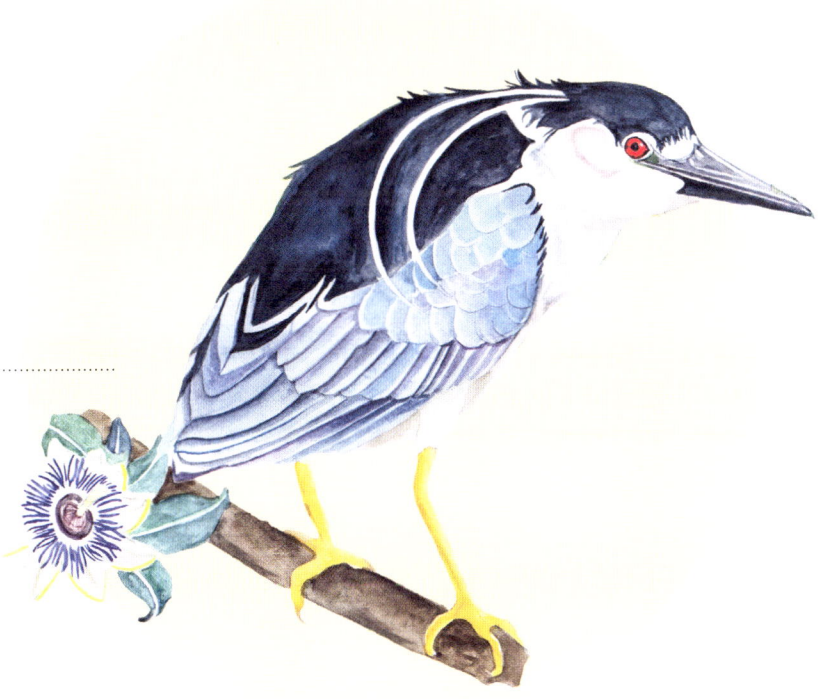

133

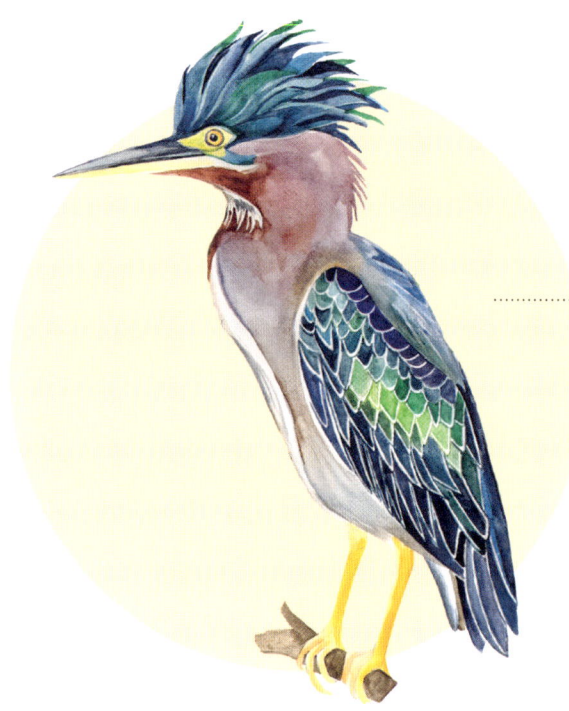

5.
GREEN HERON

Butorides virescens
Order Pelecaniformes

One of the world's only tool-using bird
species, the Green Heron creates fishing
lures using insects, feathers, and other bait
to entice fish to the water's surface.

6.
PALM WARBLER

Setophaga palmarum
Order Passeriformes

Though its name implies that it is a tropical bird,
this species breeds in Canada's boreal forests. It is
most often spotted during migration or in wintering
grounds in the Central and Southern United States.

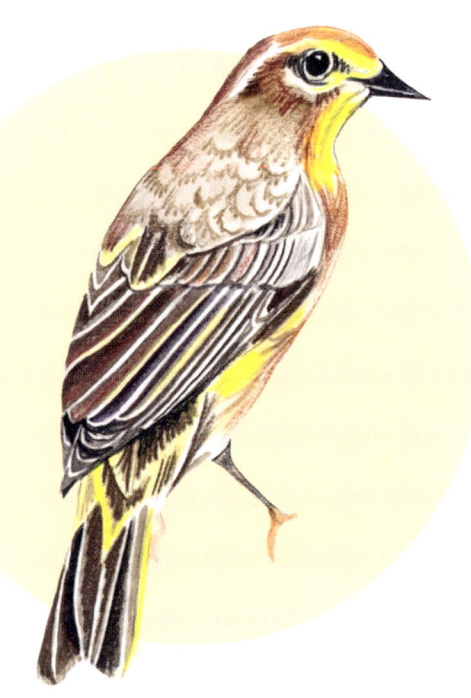

7.
PURPLE GALLINULE

Porphyrio martinica
Order Gruiformes

These large, multicolored marsh birds step across
water lilies and other floating vegetation as
they hunt for frogs and invertebrates. Pairs build
floating nests out of clumps of vegetation.

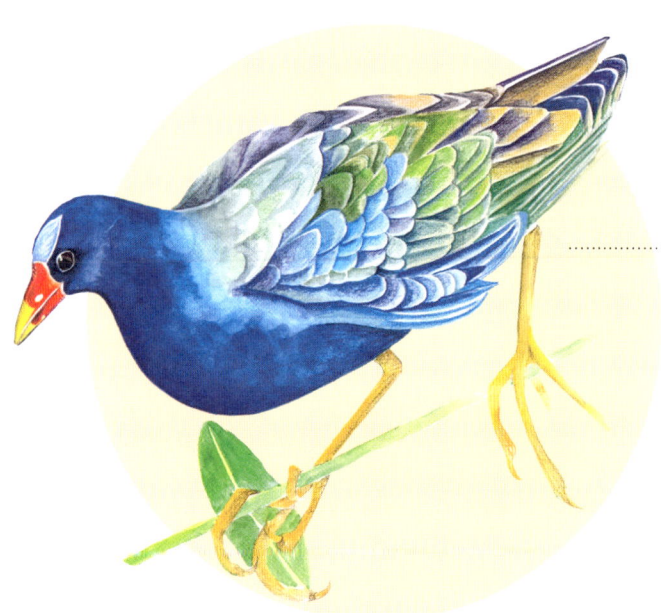

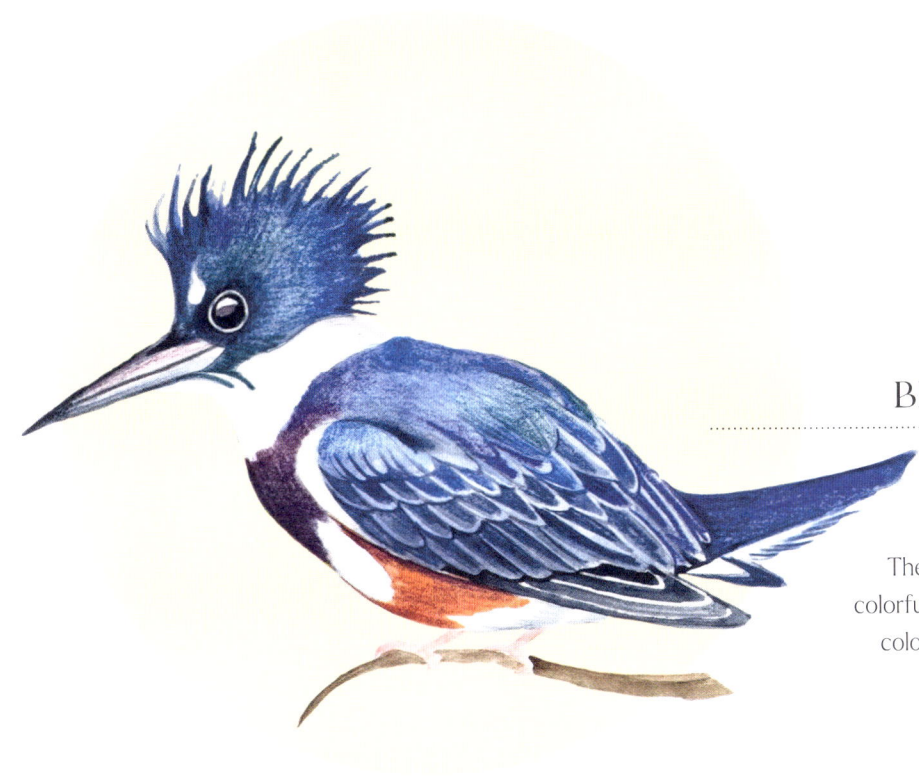

8.
BELTED KINGFISHER

Megaceryle alcyon
Order Coraciiformes

The female of this species is a bit more colorful than the male, with a distinctive rust-colored band across its chest and flanks.

9.
MANGROVE CUCKOO

Coccyzus minor
Order Cuculiformes

One of the least understood bird species of this region is highly secretive and difficult to spot, as it spends its time hidden in the most impenetrable tangles of wild mangrove forests.

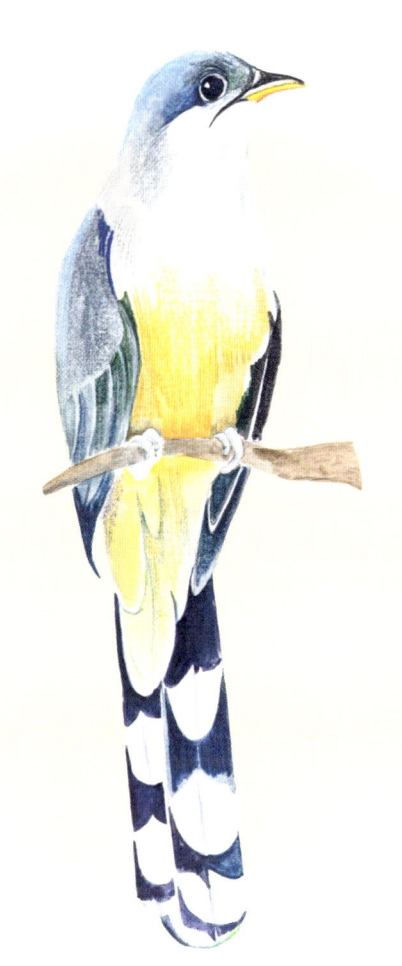

135

BIRDS
of the
SOUTHWESTERN
DESERT

Birds of the Southwest must be extremely resourceful to survive in the arid and often treeless desert habitat. Some birds are seasonal visitors and others are uniquely adapted to survive in this unforgiving climate year-round. The Cactus Wren (9), for example, rarely drinks water and instead gets its liquid needs from the insects and fruits it consumes. Same for the flame-colored Pyrrhuloxia (3), which resembles the related Northern Cardinal: it will drink from pools of water, but during the hottest and driest parts of the year, it relies on its diet of insects to stay hydrated.

Of course, any discussion of North American desert birds would be remiss if it did not include the iconic Greater Roadrunner (8), a bird so fast it can outrun a human and swiftly catch and kill a slithering rattlesnake. Several generations of American cartoon watchers know the familiar Roadrunner character from the iconic Saturday morning cartoon. Contrary to his depiction in the cartoon, however, his nemesis, Wile E. Coyote, can run at tremendous speeds—nearly twice as fast as those of a roadrunner. As a result, coyotes pose real danger to these birds. Greater Roadrunners hold symbolic significance in Native American myth and legend. Their distinctive X-shaped footprints—with two toes pointing forward and two backward—disguise which direction the birds are running and are thought by the Pueblo people to hinder evil spirits from following.

Another significant bird is the fearsome, red-eyed Cooper's Hawk (2). You may catch a glimpse of this raptor soaring above the desert in search of prey. In bird divination, the practice of deriving meaning from the behaviors of birds, this agile hawk is believed to represent strength and power in the face of adversity.

1.	3.	5.	7.	9.
VERDIN	PYRRHULOXIA	ANNA'S HUMMINGBIRD	LARK SPARROW	CACTUS WREN
2.	4.	6.	8.	10.
COOPER'S HAWK	CURVE-BILLED THRASHER	GILA WOODPECKER	GREATER ROADRUNNER	GAMBEL'S QUAIL

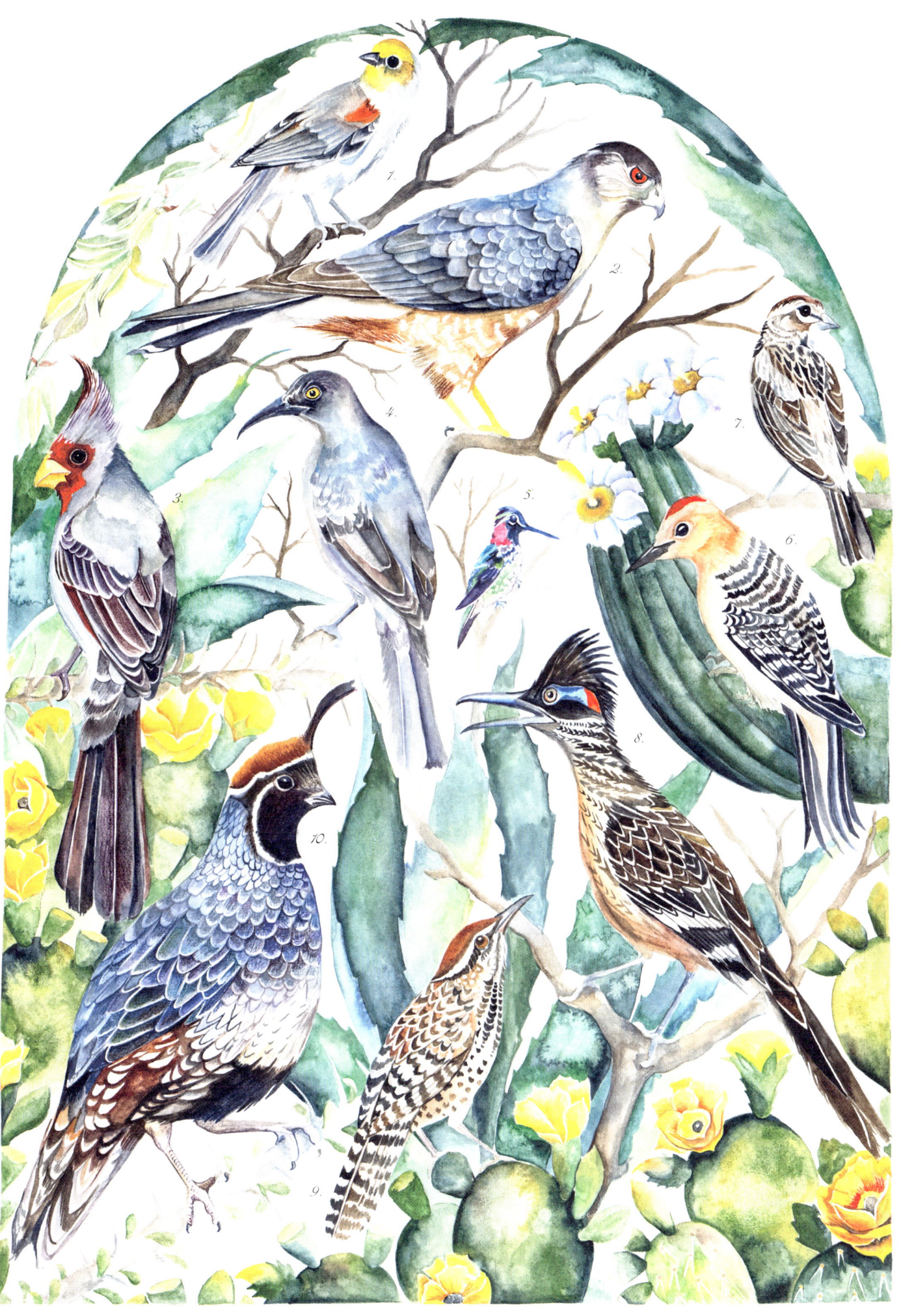

1.
VERDIN

Auriparus flaviceps
Order Passeriformes

These songbirds feed primarily on insects, but they supplement their diets with fruit, seeds, and flower nectar, sometimes drinking from hummingbird feeders when flower nectars are scarce.

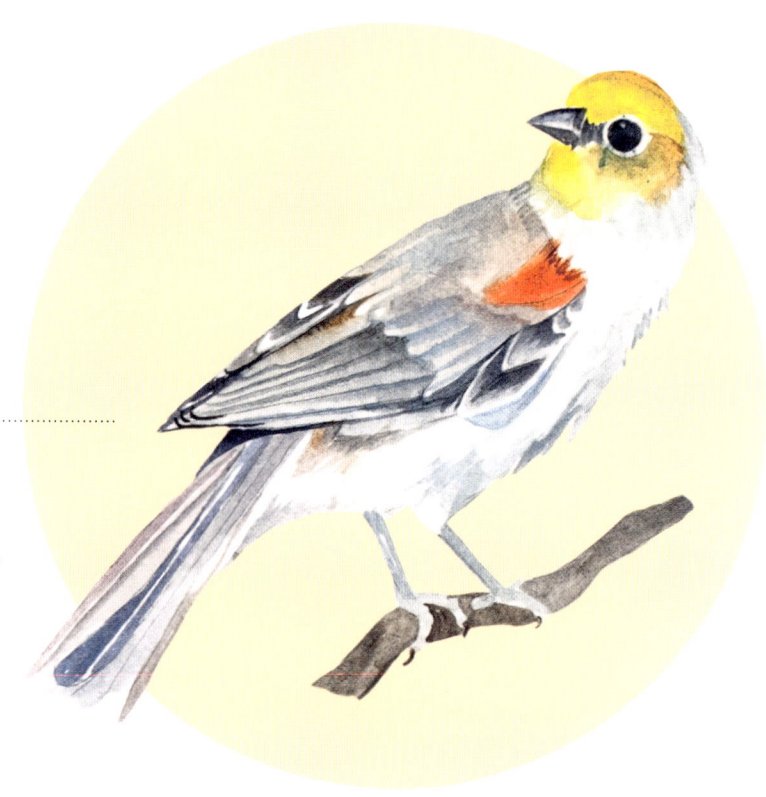

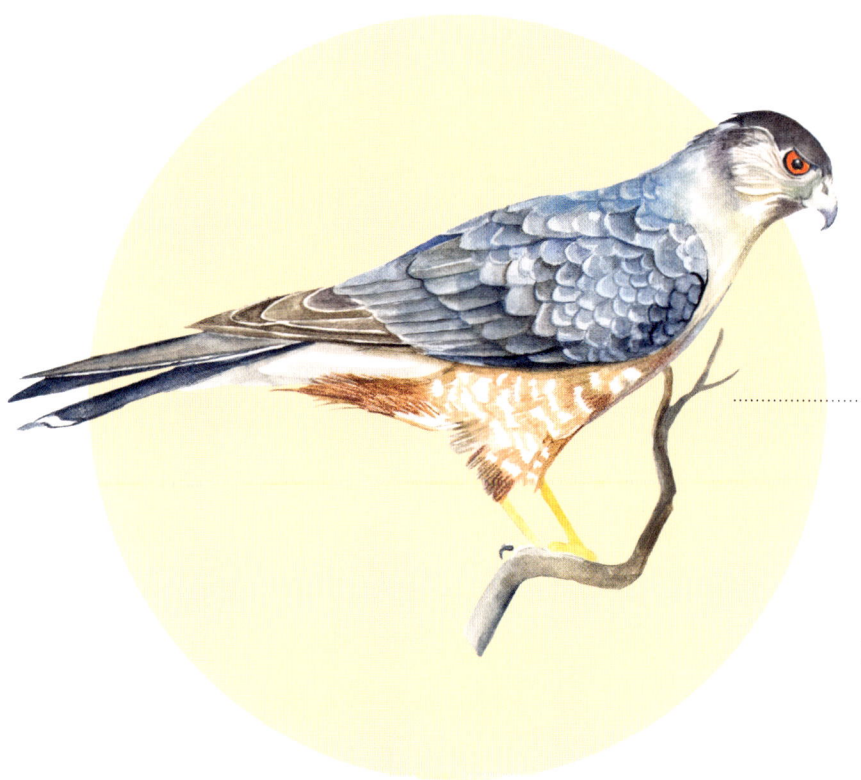

2.
COOPER'S HAWK

Accipiter cooperii
Order Accipitriformes

These common hawks are among the world's most skillful fliers as they tear through tree canopies at high speeds in pursuit of other birds. They sometimes hang around at backyard birdfeeders, looking for an easy meal.

3.
PYRRHULOXIA

Cardinalis sinuatus
Order Passeriformes

This tough songbird's common name comes from two Latin words that mean "flame-colored" and "crooked," which aptly describe this reddish bird with a crooked bill.

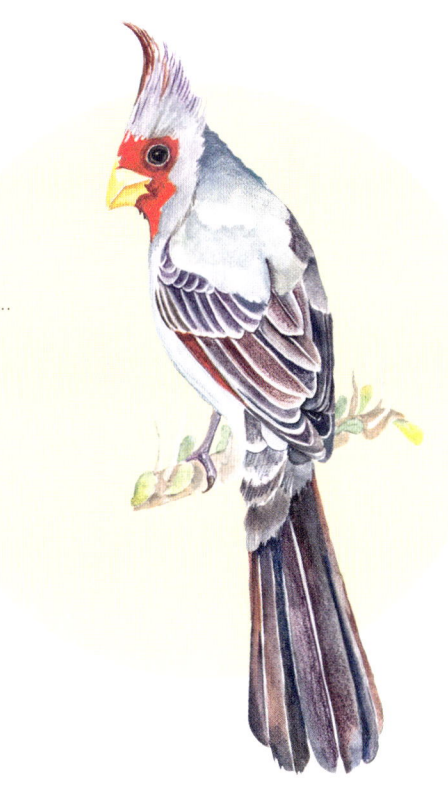

4.
CURVE-BILLED THRASHER

Toxostoma curvirostre
Order Passeriformes

The long, curved bill that is characteristic of this species helps it forage for insects among the prickly plants and cacti of the desert.

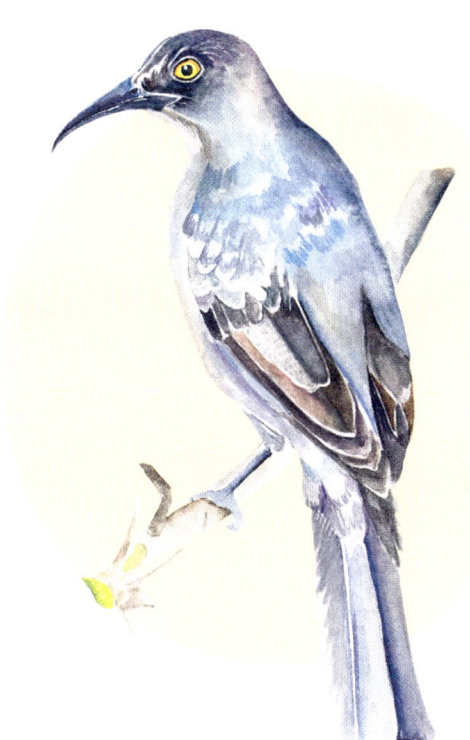

5.
ANNA'S HUMMINGBIRD

Calypte anna
Order Caprimulgiformes

Although it is found year-round in the Southwest, this hummer is particularly common and breeds in the Northwest, where it is the official bird of Vancouver, British Columbia.

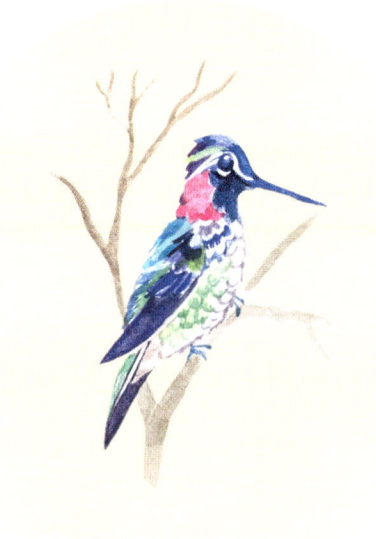

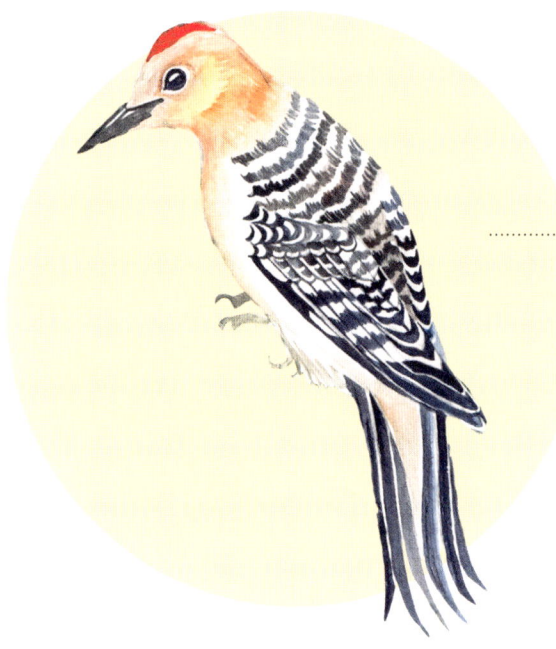

6.
GILA WOODPECKER

Melanerpes uropygialis
Order Piciformes

These birds excavate nest cavities in saguaro cacti, which, once vacated by the woodpecker, are often taken over by other birds, including owls and Cactus Wrens.

7.
LARK SPARROW

Chondestes grammacus
Order Passeriformes

Female Lark Sparrows occasionally use old nests of mockingbirds or thrashers instead of building their own.

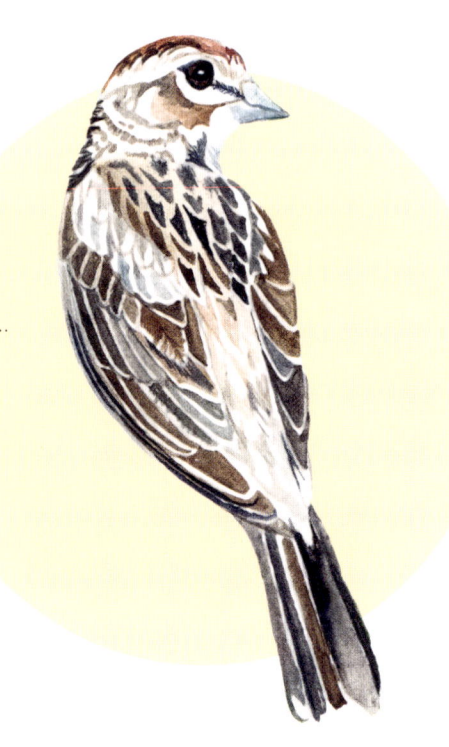

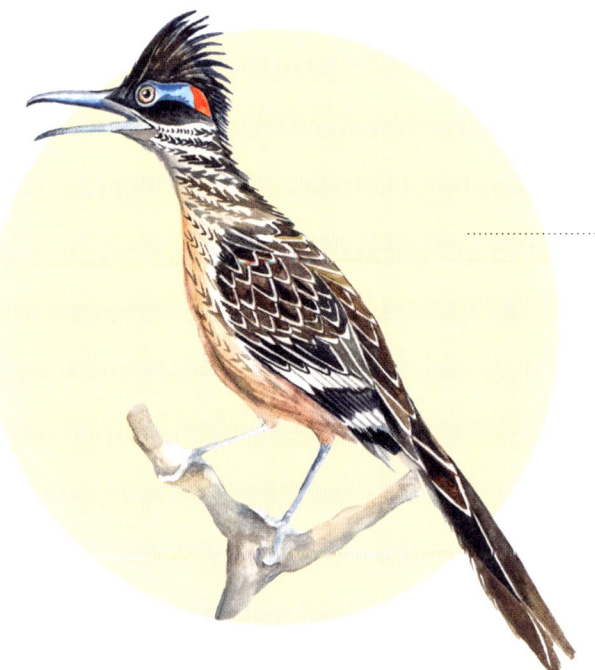

8.
GREATER ROADRUNNER

Geococcyx californianus
Order Cuculiformes

Roadrunners eat poisonous lizards and deadly scorpions without ill effects. A roadrunner will often work with another roadrunner to distract a rattlesnake while its partner grabs the snake's head and bashes it against a rock before eating it.

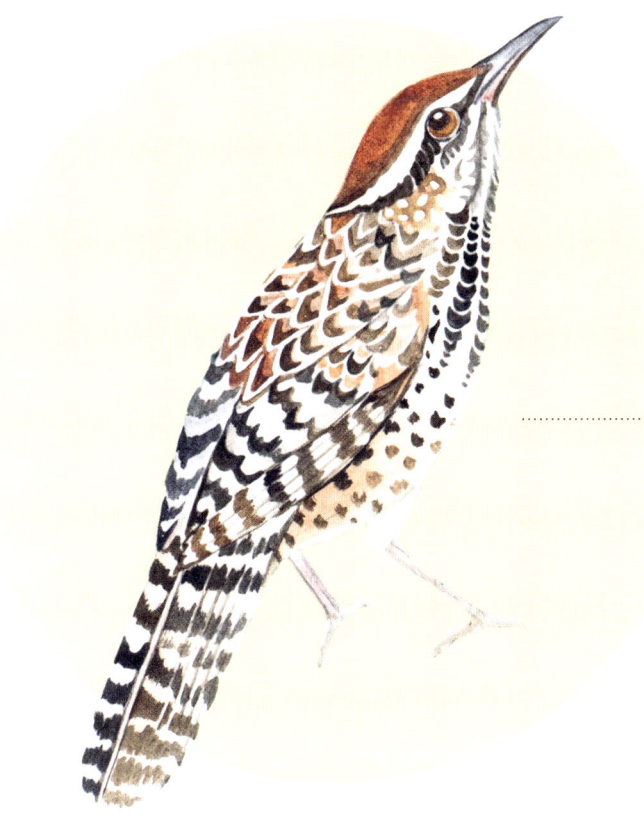

9.
CACTUS WREN

Campylorhynchus brunneicapillus
Order Passeriformes

This noisy desert dweller, the largest wren species
in the United States, is the state bird of Arizona.
They sometimes destroy the nests of other bird
species, pecking or removing their eggs.

10.
GAMBEL'S QUAIL

Callipepla gambelii
Order Galliformes

These gregarious birds have adapted to reproduce
in relationship to seasonal changes. A year with
more winter–spring rainfall will generally result
in more food and a greater number of chicks.

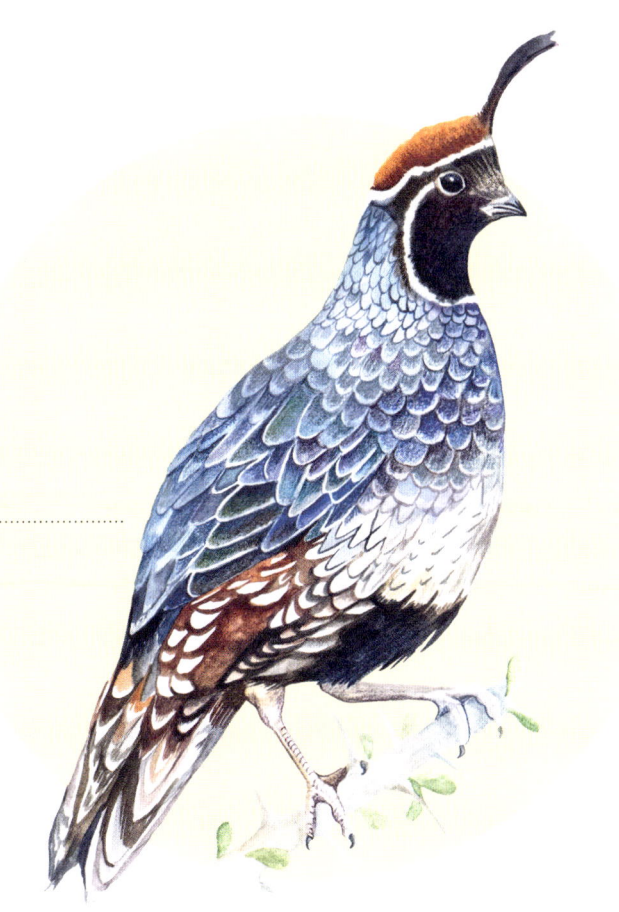

BIRDS
of the
NORTHWEST

In the rugged mountain forests of the Northwest, along slopes thick with wildflowers and glassy lakes that mirror soaring, ancient trees, keep your eyes open and your ears alert and you may see and hear some of the magnificent birds that inhabit or visit the region. One of the most notable songsters here is the Northern Mockingbird (3). The state bird of five US states, the mockingbird's beautiful song has inspired numerous works of music and literature and has become a potent symbol of beauty and innocence. The mockingbird appears in the mythology and traditions of several Native American cultures. In the creation myths of several Pueblo tribes, it teaches the first people to speak. In several other tribal traditions, it is considered a harbinger of wisdom, good fortune, and supernatural powers.

The melodious call of the mockingbird may be overshadowed by another Northwest bird, however. The melancholy song of the Hermit Thrush (2) is considered one of the most remarkable of all songbirds in North America. These shy, mellifluous songbirds are difficult to spot, but in the evening, as you sit beneath the trees and dusk settles in around the forest, you may be serenaded by their lilting, sweet song.

The Steller's Jay (9) is a common sight in the Northwest, a striking blue bird capable of mimicking various sounds, including those of other birds and animals. Like Blue Jays, Steller's Jays are bullies. Highly territorial and aggressive, these opportunistic omnivores are known to attack the nests of other birds in their vicinity if they feel threatened or to steal and eat eggs or nestlings.

If you see a flash of blue in the trees, don't immediately assume it's a Steller's Jay; it may be a Lazuli Bunting (4). Interestingly, the vibrant blue of this songbird—and that of any other bird—is not actually blue feather pigment, but the structure of its delicate feathers that diffract light, making them appear blue to us.

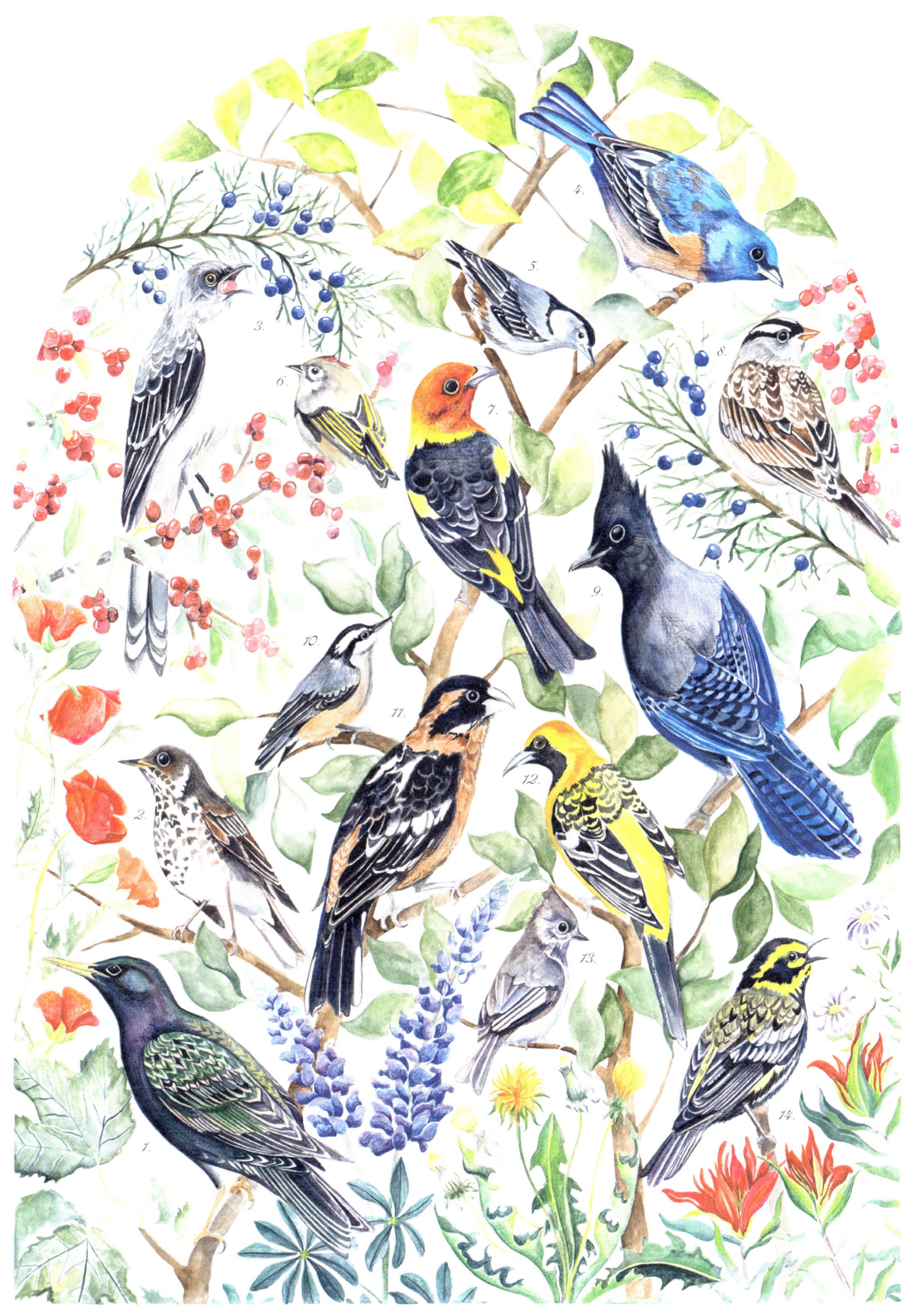

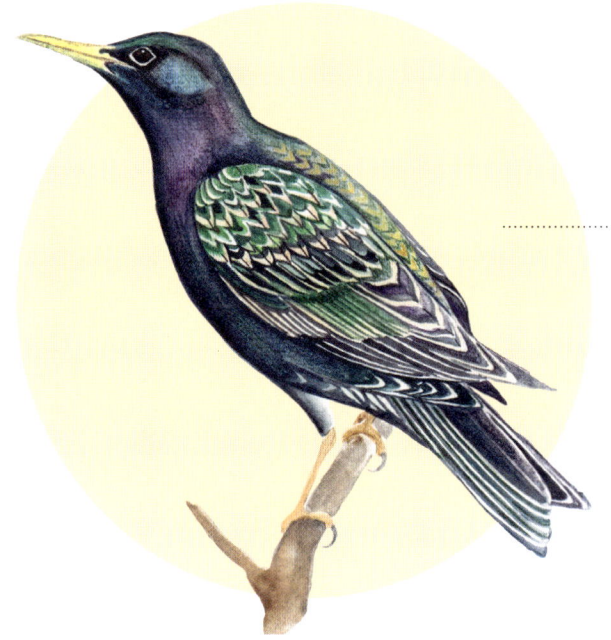

1.
EUROPEAN STARLING

Sturnus vulgaris
Order Passeriformes

Starlings can mimic the calls of up to twenty different species, including the American Robin, the Killdeer, the Northern Bobwhite, the Northern Flicker, and the Wood Thrush.

2.
HERMIT THRUSH

Catharus guttatus
Order Passeriformes

These songbirds are found throughout the United States. Those that live in the West tend to build their nests in trees and shrubs, while their eastern brethren prefer building nests on the ground.

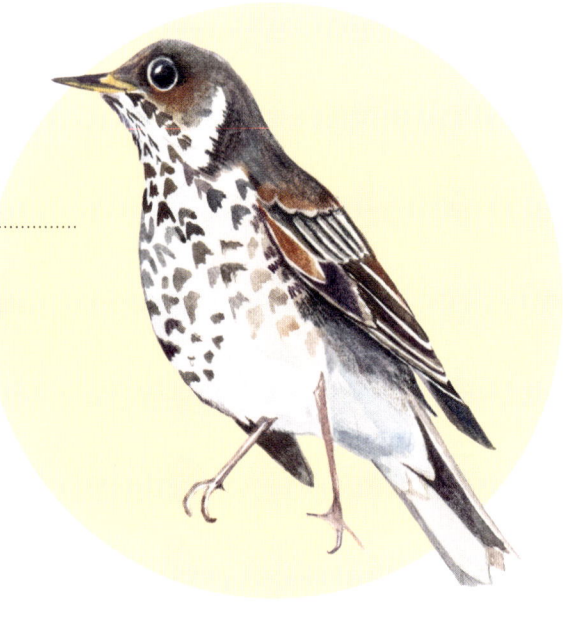

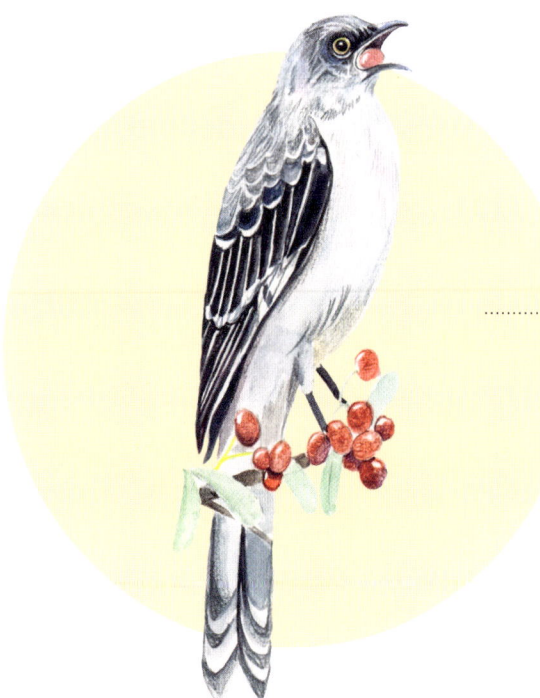

3.
NORTHERN MOCKINGBIRD

Mimus polyglottos
Order Passeriformes

This songbird can string ten to fifteen different songs in an endless loop, sometimes even at night.

4.
LAZULI BUNTING

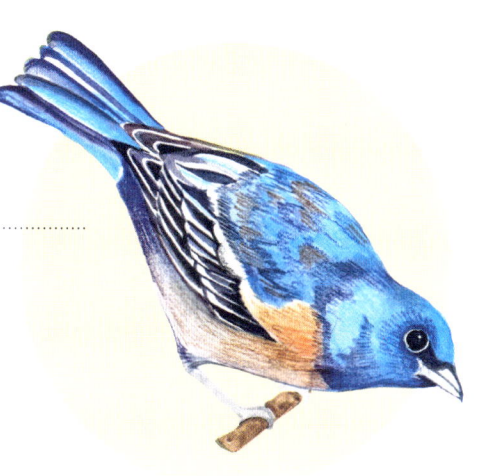

Passerina amoena
Order Passeriformes

Each male of this species sings a unique combination of notes that it creates as a yearling. It then sings the same song for the rest of its life.

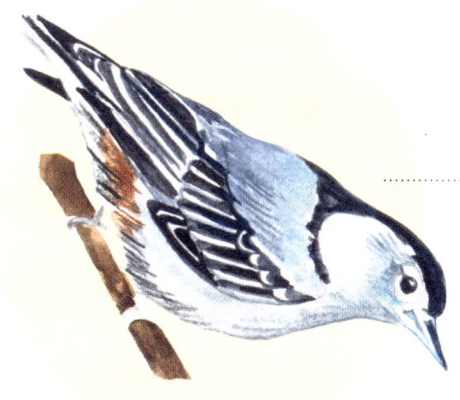

5.
WHITE-BREASTED NUTHATCH

Sitta carolinensis
Order Passeriformes

Known to form scavenging flocks with other species such as chickadees and titmice, these birds forage for food at birdfeeders and in open areas.

6.
RUBY-CROWNED KINGLET

Corthylio calendula
Order Passeriformes

The jumbled song of this tiny bird, an ascending, accelerating series of very high-pitched *tsee* notes, is much louder than its tiny body might suggest.

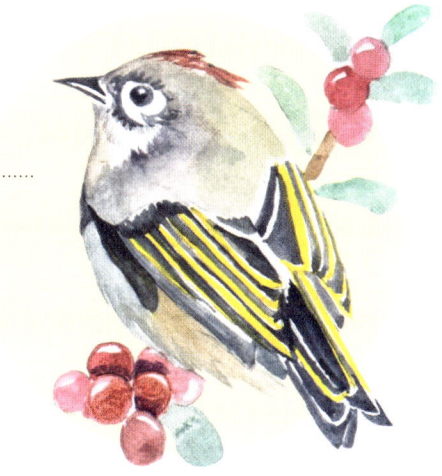

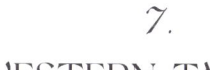

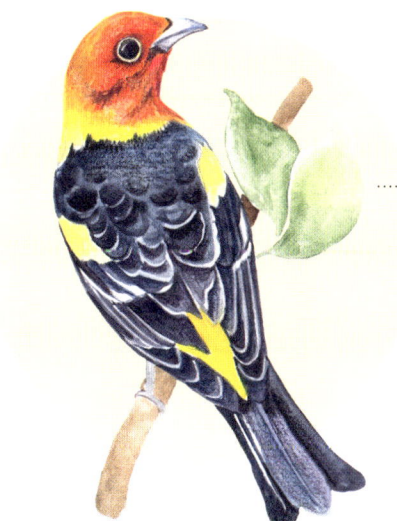

7.
WESTERN TANAGER

Piranga ludoviciana
Order Passeriformes

Most red birds' feather colors are derived from carotenoids in their diets, but this species' scarlet head feathers are from a rare pigment called rhodoxanthin, which probably comes from the insects they eat.

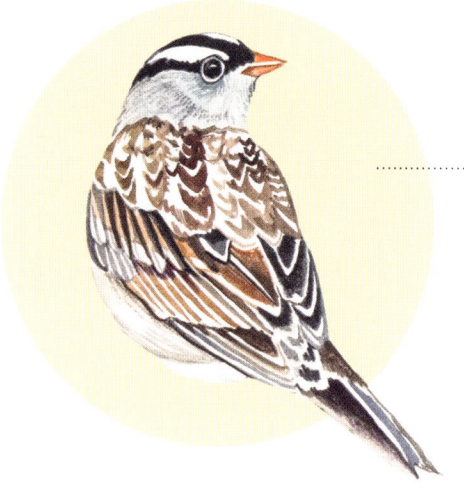

WHITE-CROWNED SPARROW

Zonotrichia leucophyrs
Order Passeriformes

The thin, sweet song of this sparrow is largely dictated by the environment where it hatches. Each has a unique song, depending on where it spent the first two or three months of its life.

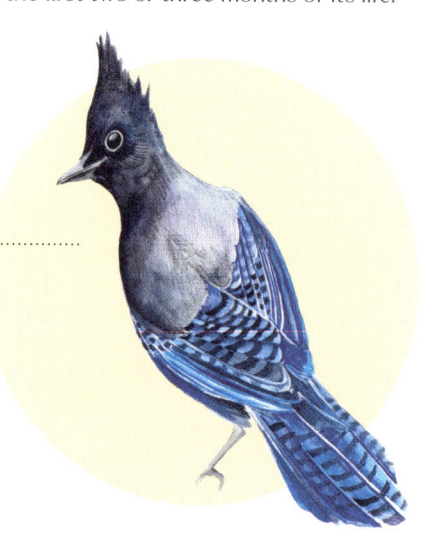

9.
STELLER'S JAY

Cyanocitta stelleri
Order Passeriformes

This species' appearance varies throughout its range, with some birds displaying black crests and backs, while others are blue.

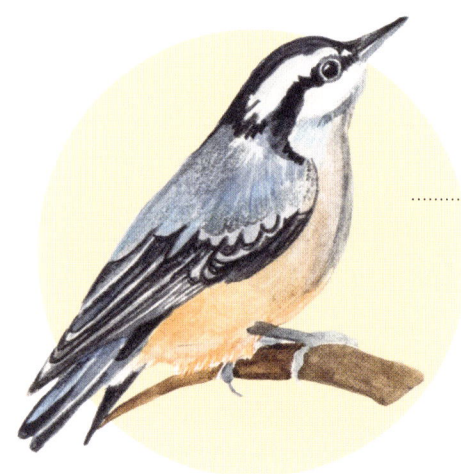

10.
RED-BREASTED NUTHATCH

Sitta canadensis
Order Passeriformes

After collecting globs of resin from conifers, this bird plasters it around the entrance to its nest hole using a piece of bark as an applicator. It dives directly through the hole to avoid the resin and reach its nest.

11.
BLACK-HEADED GROSBEAK

Pheucticus melanocephalus
Order Passeriformes

Males and female grosbeaks are loud singers, with the female's song being a simpler version of the male's tune. Occasionally, a female will sing a full male song, possibly to deceive its mate about the presence of intruders to convince him to spend more time at the nest.

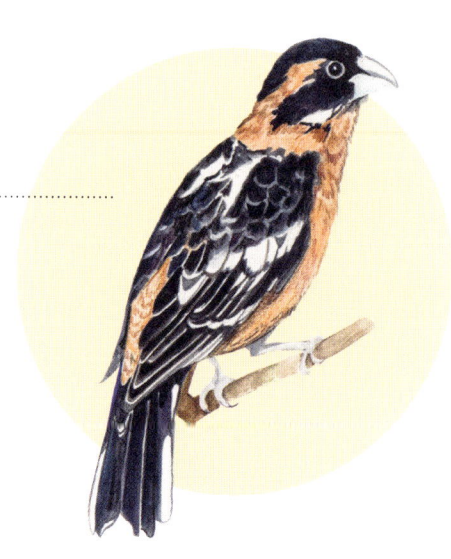

12.
HOODED ORIOLE

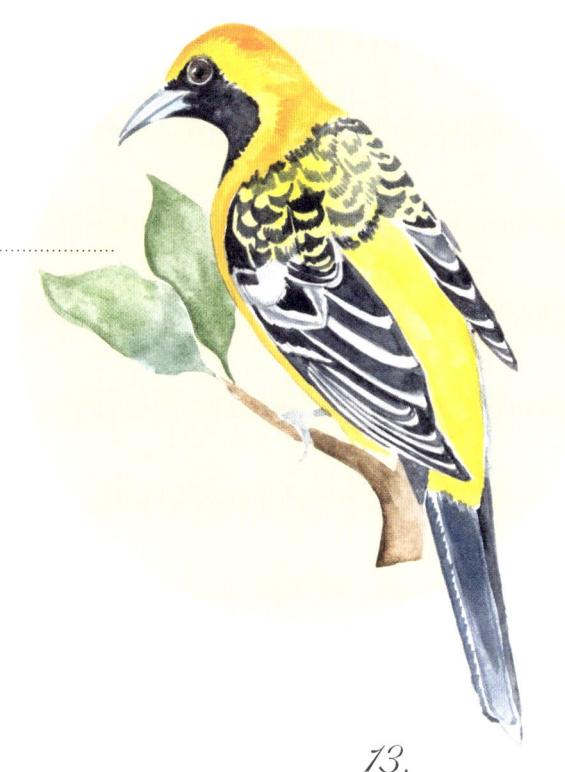

Icterus cucullatus
Order Passeriformes

Although this species is primarily a resident of the Southwest and California, they sometimes range into the Pacific Northwest as a "spring overshoot" of migrants returning from Mexico.

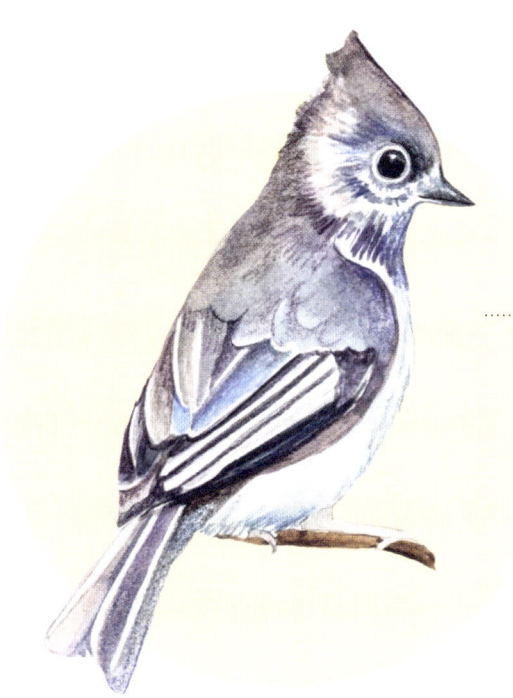

13.
TUFTED TITMOUSE

Baeolophus bicolor
Order Passeriformes

When a titmouse finds a large seed at a feeder, it carries it to a safer, more secluded perch, cracks it with a few sharp whacks of its bill, and then consumes it or hides it to eat later.

14.
TOWNSEND'S WARBLER

Setophaga townsendi
Order Passeriformes

This species sometimes begins constructing its nest in one tree, but then changes its mind and moves everything to another tree to finish the nest there.

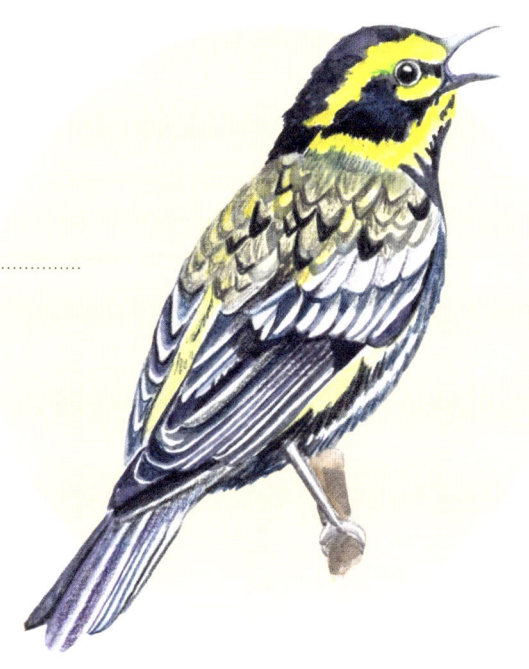

BACKYARD BIRDS of NEW ENGLAND

Many beautiful birds pass through backyards and garden birdfeeders in rural New England. On a sunny spring day, you may spy an Eastern Bluebird (2). For many, catching sight of this cheerful little bird is the first sign that the coldest winter days are in the past and spring has arrived. American colonists called this bird the "blue robin," given its resemblance to the European Robin. Bluebirds are a ubiquitous presence in the works of many poets and songwriters and have inspired numerous popular songs and works of literature. They also hold mythological significance in several cultures, reappearing as symbols of hope, happiness, and cheer.

If your birdwatching takes you to the forest edge, you may hear the melodious voice of the Wood Thrush (6). According to Irish folklore, if a thrush builds its nest close to the ground, it is auspicious for the fairies, as they will be able to hear his beautiful music, but if it builds its nest high up, the fairies will become enraged and bring misfortune upon the inhabitants of the area.

Meanwhile, the ubiquitous Barn Swallow (7)—easy to spot swooping in and out of barns and other structures—is said in some legends to have a forked tail as punishment for stealing fire from the gods to bring to humans. In frustration, the gods threw a ball of fire at the bird, burning off its middle tailfeathers and leaving the forked tail that distinguishes it.

The first time you see a Rose-breasted Grosbeak (12), you could be forgiven for thinking you are hallucinating a tropical jungle bird—because in some ways, you'd be right. Recognizable by its stout bill and the triangular red patch of plumage on its otherwise black-and-white body, this species migrates as far south as South America, fueled by the fruits it consumed in the north.

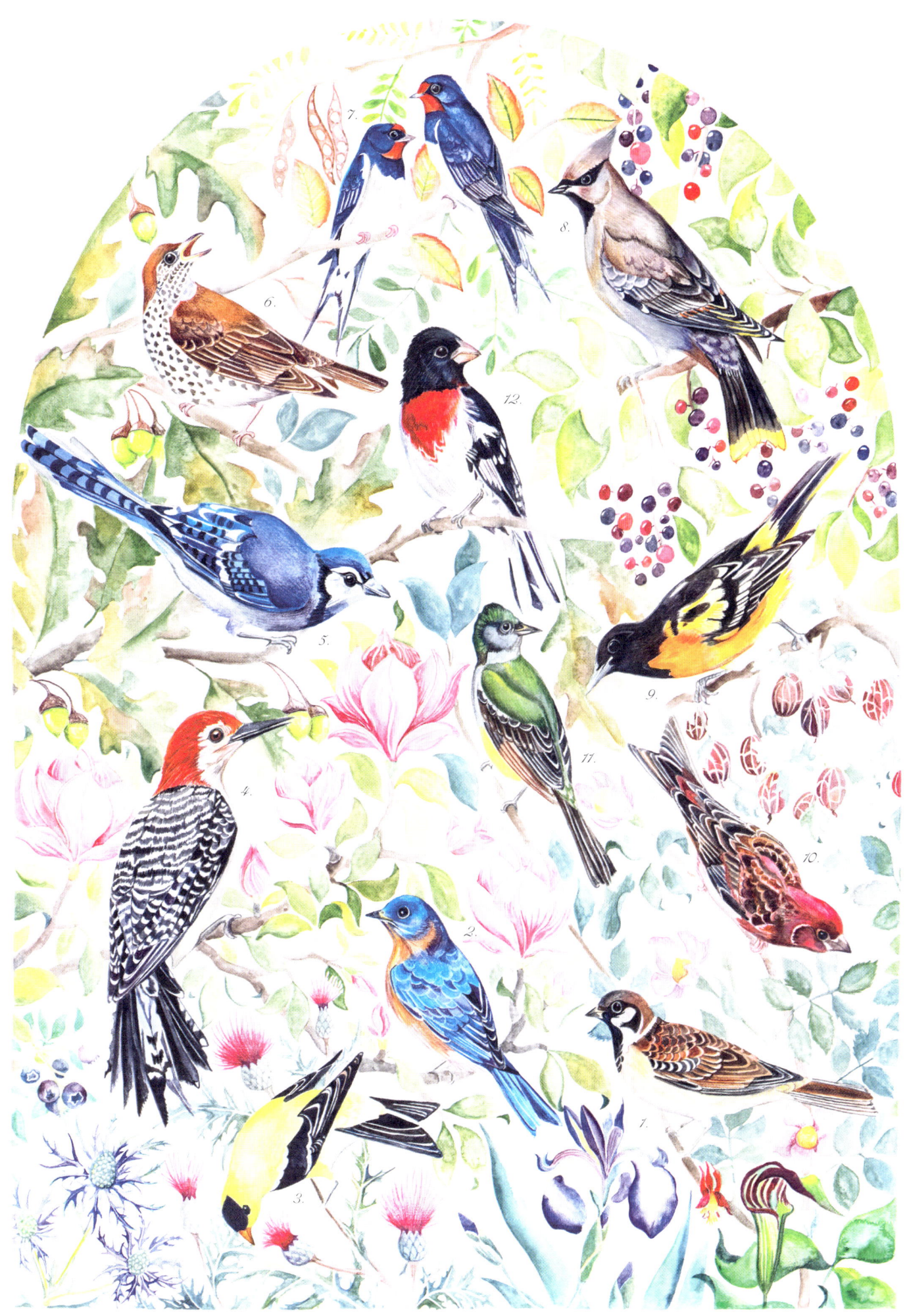

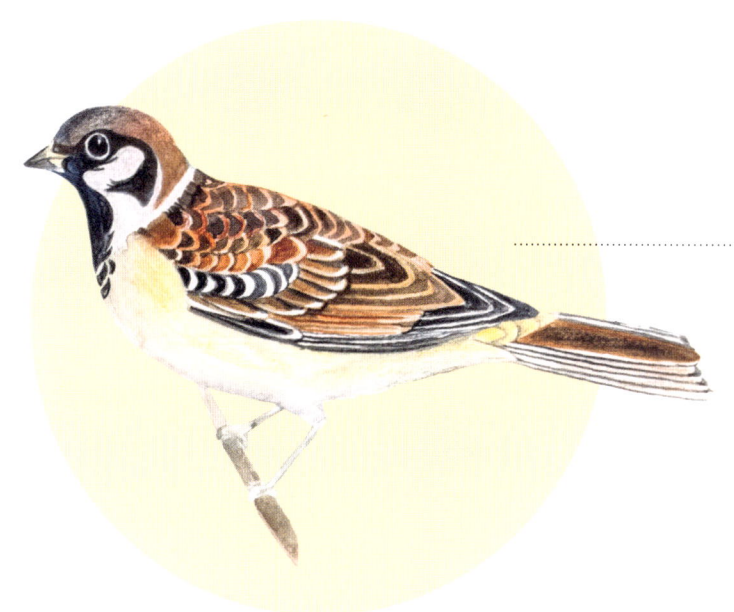

1.
HOUSE SPARROW

Passer domesticus
Order Passeriformes

This ubiquitous species doesn't like to build a nest in a tree. Instead, it prefers to nest in the eaves of buildings, in walls, on utility poles, and on other human-made structures.

2.
EASTERN BLUEBIRD

Sialia sialis
Order Passeriformes

This species' blue color is the result of refracted light rather than pigmentation. Its feathers are filled with tiny air pockets that reflect sunlight in such a way that only blue wavelengths are reflected, making them appear blue to us.

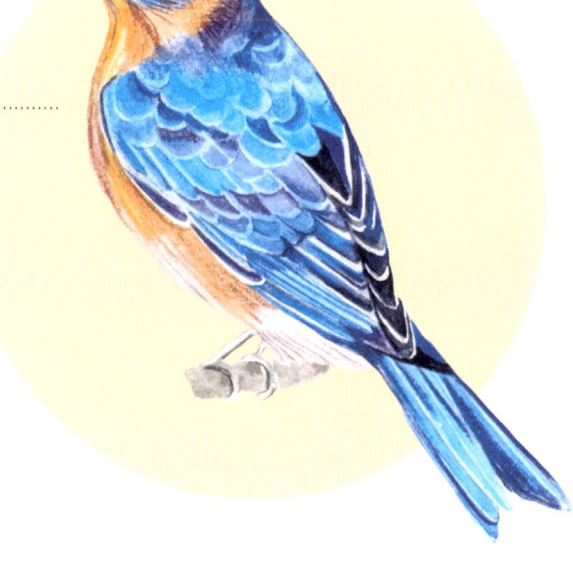

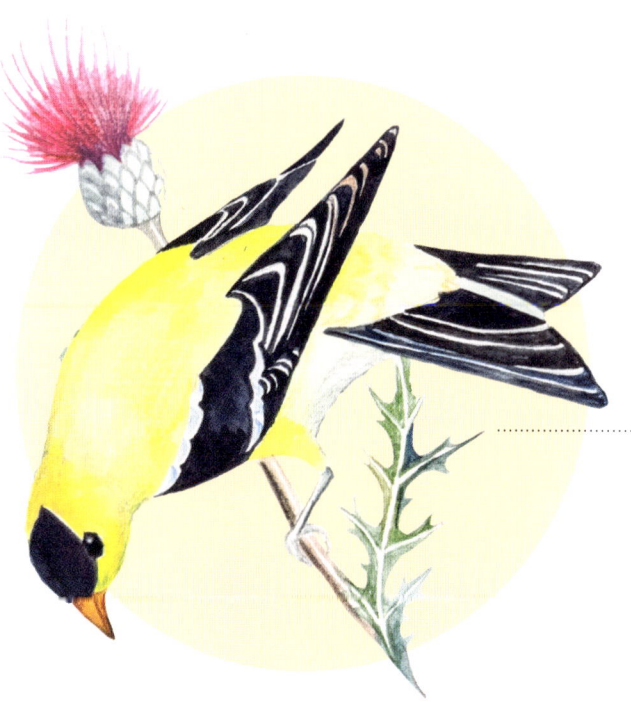

3.
AMERICAN GOLDFINCH

Spinus tristis
Order Passeriformes

The goldfinch is one of very few birds in the world whose diet is entirely vegetarian, including seeds, flowers, and other plants.

4.
RED-BELLIED WOODPECKER

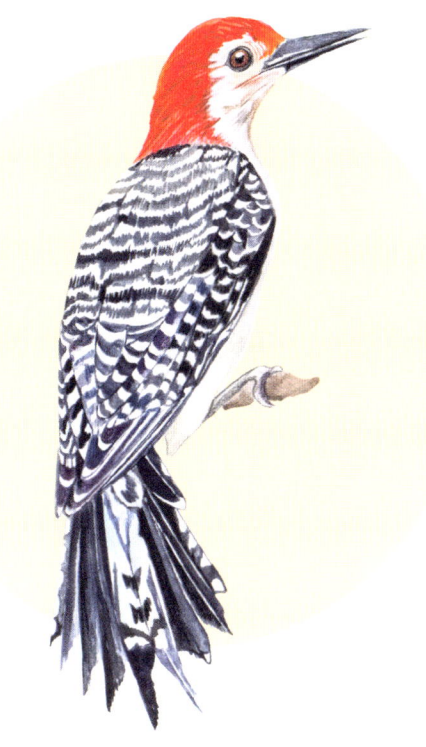

Melanerpes carolinus
Order Piciformes

These opportunistic nesters will sometimes take over the nests of other birds, including smaller species of woodpecker. On the other hand, more than half of this species' nests are invaded by aggressive European Starlings.

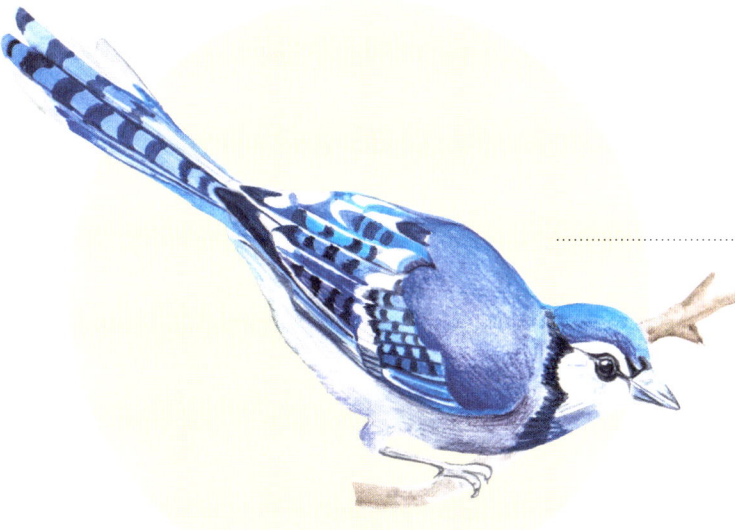

5.
BLUE JAY

Cyanocitta cristata
Order Passeriformes

Blue Jays are known as bullies and troublemakers in the bird world. They are very territorial and sometimes eat the eggs and nestlings of other birds.

6.
WOOD THRUSH

Hylocichla mustelina
Order Passeriformes

Known for its beautiful song, a Wood Thrush can produce two harmonizing notes simultaneously.

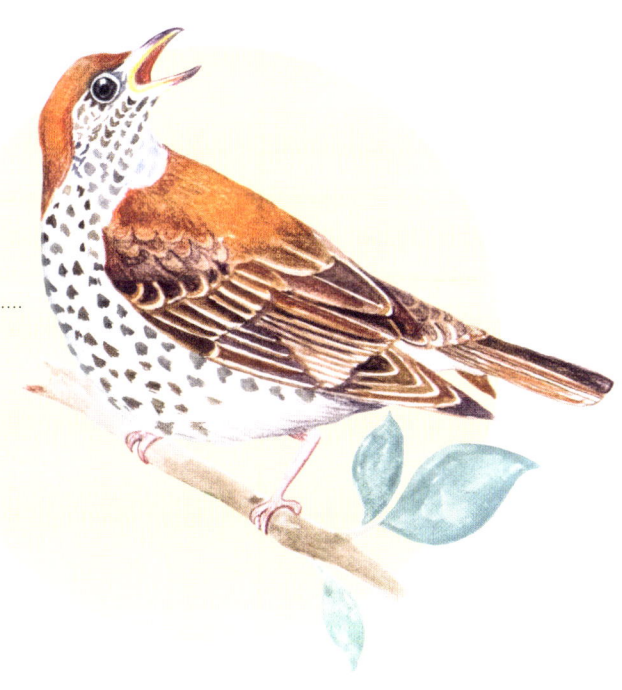

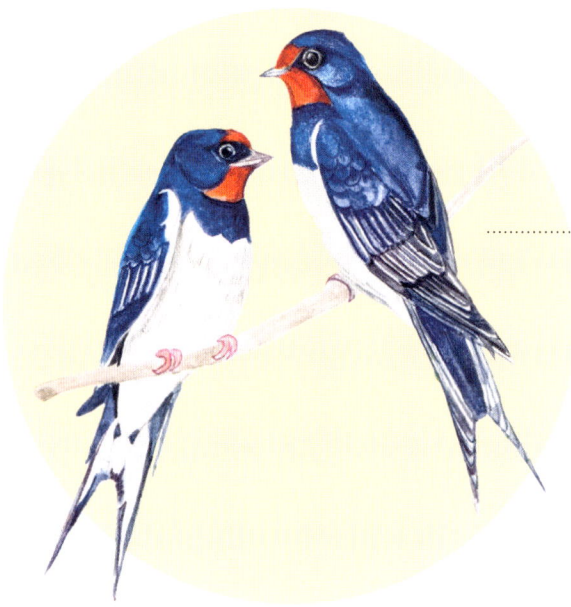

7.
BARN SWALLOW

Hirundo rustica
Order Passeriformes

These birds are named for their preference for nesting in barns and other human-made structures. It is common to see them swooping in and out of the rafters of barns.

8.
CEDAR WAXWING

Bombycilla cedrorum
Order Passeriformes

The yellow tip of this species' tail is a result of pigment in the berries it eats. Slight variations and more orange-toned tail tips result when a bird eats more honeysuckle berries.

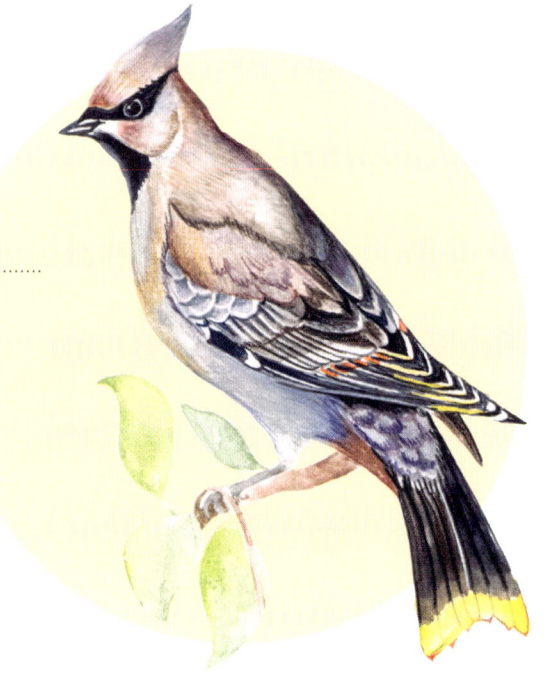

9.
BALTIMORE ORIOLE

Icterus galbula
Order Passeriformes

Rather than snacking on underripe fruit like many other birds, Baltimore Orioles tend to seek out the very ripest and darkest berries.

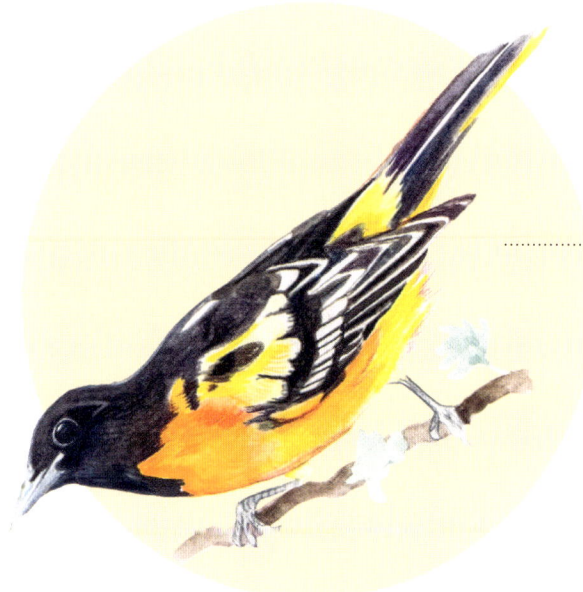

10.
PURPLE FINCH

Haemorhous purpureus
Order Passeriformes

Noted American ornithologist Roger Tory Peterson famously described this bird as "a sparrow dipped in raspberry juice."

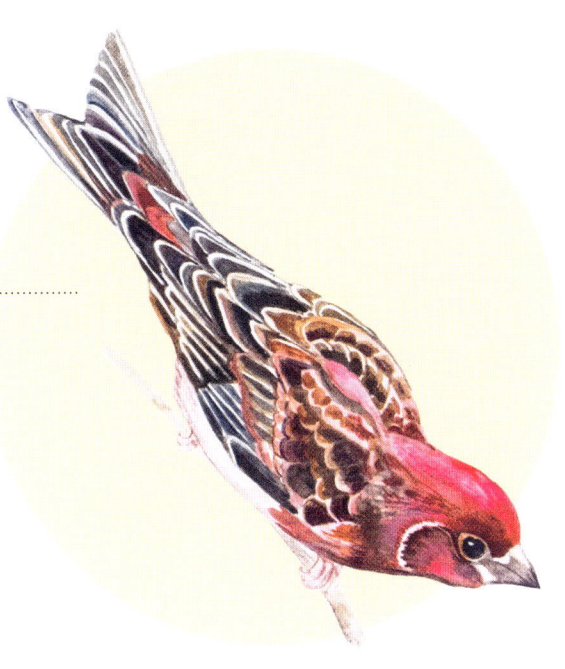

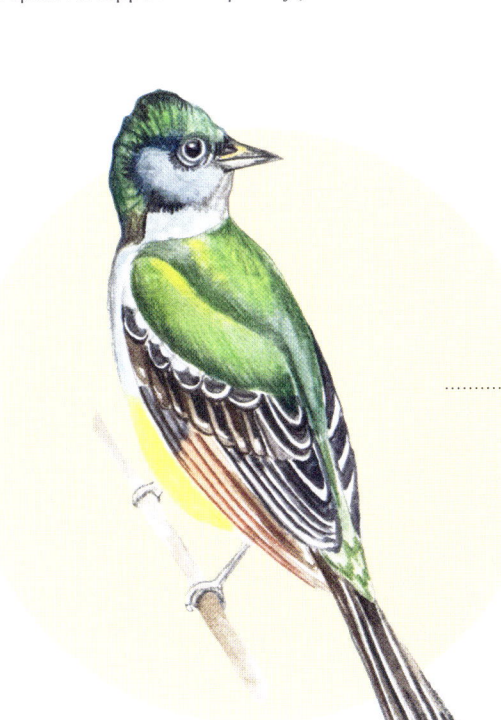

11.
GREAT CRESTED FLYCATCHER

Myiarchus crinitus
Order Passeriformes

This creative nest builder weaves bits of shed snakeskin and other delicate found materials such as cellophane wrappers and bits of paper into its nest.

12.
ROSE-BREASTED GROSBEAK

Pheucticus ludovicianus
Order Passeriformes

This bird is renowned for its exceptionally beautiful, robin-like song. Some suggest it sounds like an opera singer.

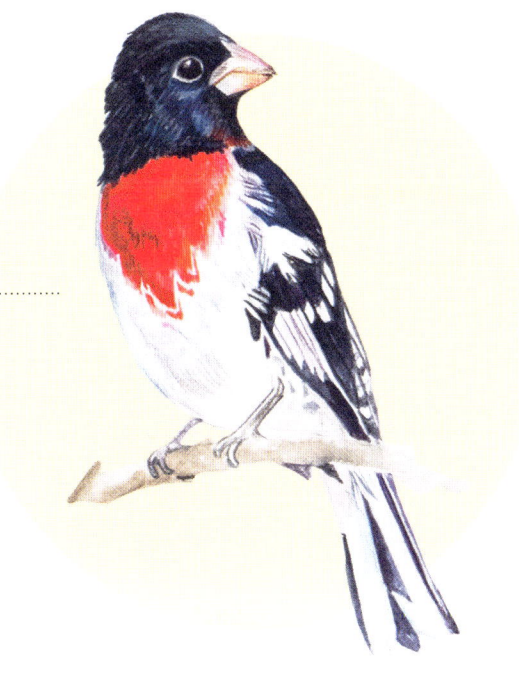

153

BIRDS
of the
SOUTH CENTRAL
UNITED STATES

The rolling hills, meandering forests, velvety blue mountains, and flowering open meadows of the South Central United States are home to a variety of birds, some of which are quite common and easy to spot throughout the backcountry.

One of the most familiar birds in this region, and throughout North America, is the Red-winged Blackbird (1), a common visitor to wetlands, swamps, and flooded roadsides. It often perches on cattails, fenceposts, and telephone wires. The scarlet markings on its shoulders are particularly eye-catching in flight. According to Native American folklore, the bird gained its signature red wings when an angry man set a marsh on fire and a blackbird came to warn the local people of the danger. The man threw a rock at the bird to silence it, and its wings began to bleed from the injury. Henceforth, the bird's wings bear two red marks.

Another common bird, the American Robin (2), is a cheerful, red-breasted songbird that hops across lawns and parks in search of food. Considered a near universal symbol of spring, robins have appeared in various songs and poems, usually as emblems of hope, joy, and positivity. The distinctive turquoise-blue egg color has led to human-made hues and variations, commonly referred to as "robin's egg blue."

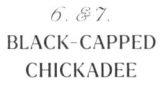

1. RED-WINGED BLACKBIRD	*4.* NORTHERN FLICKER	*6. & 7.* BLACK-CAPPED CHICKADEE	*9.* NORTHERN CARDINAL	*11.* RUBY-THROATED HUMMINGBIRD
2. AMERICAN ROBIN	*5.* EASTERN MEADOWLARK	*8.* INDIGO BUNTING	*10.* EVENING GROSBEAK	*12.* ORCHARD ORIOLE
3. SCARLET TANAGER				*13.* COMMON GRACKLE

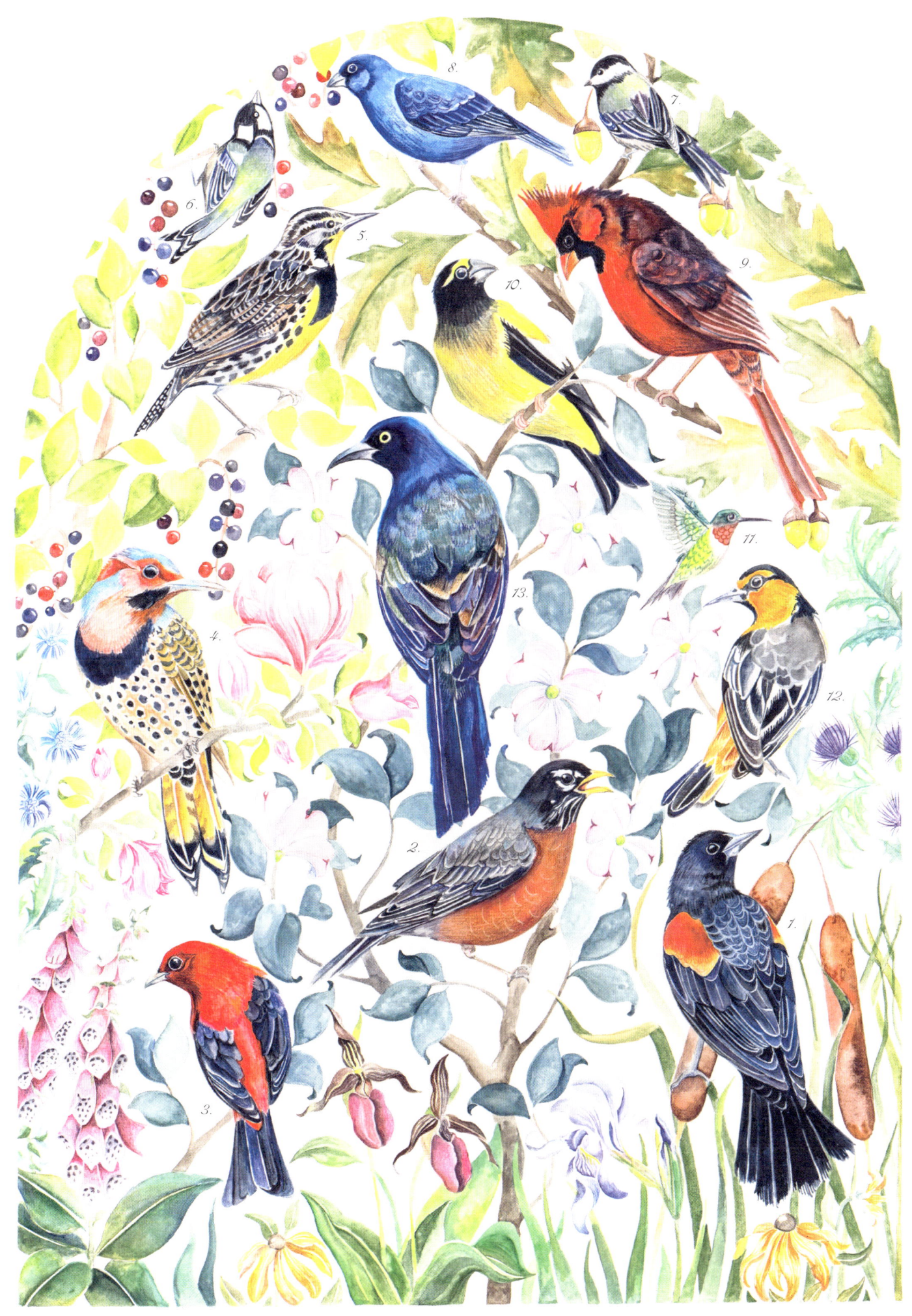

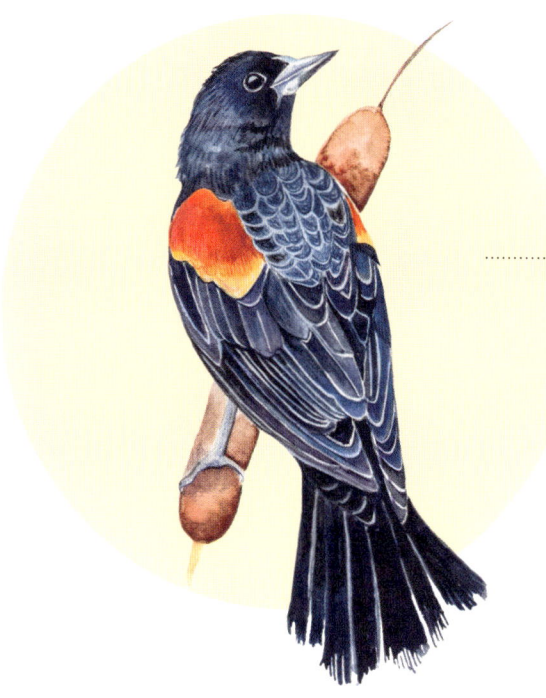

1.
RED-WINGED BLACKBIRD

Agelaius phoeniceus
Order Passeriformes

The Red-winged Blackbird is a polygamous species. Males may have as many as fifteen nesting female mates with whom they breed.

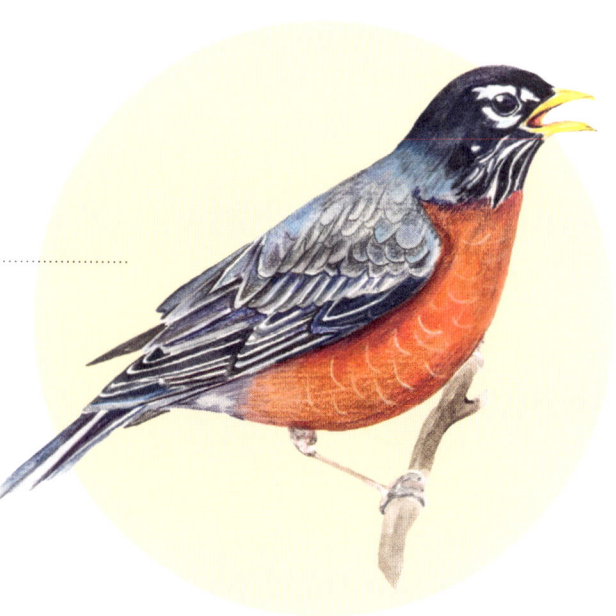

2.
AMERICAN ROBIN

Turdus migratorius
Order Passeriformes

This ubiquitous North American bird eats a lot of fruit in fall and winter. If they eat too many honeysuckle berries, they can become intoxicated.

3.
SCARLET TANAGER

Piranga olivacea
Order Passeriformes

This bird is named for the male's scarlet plumage during breeding season. When breeding season is over, the males molt to resemble females, and their red feathers are replaced by yellow-green plumage.

4.
NORTHERN FLICKER

Colaptes auratus
Order Piciformes

Like other woodpeckers, flickers usually nest in holes in trees, but they also nest in burrows vacated by Belted Kingfishers or Bank Swallows.

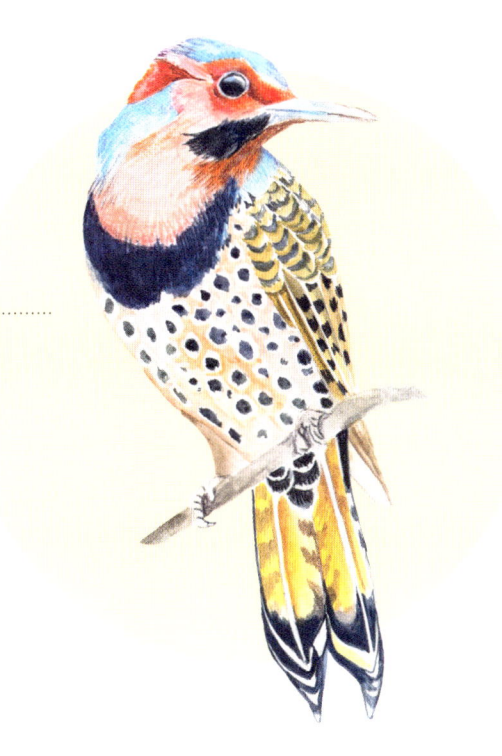

5.
EASTERN MEADOWLARK

Sturnella magna
Order Passeriformes

Although Eastern and Western meadowlarks are nearly identical, the two species almost never mix. In fact, where their habitats intersect, they sometimes engage in battles for territory.

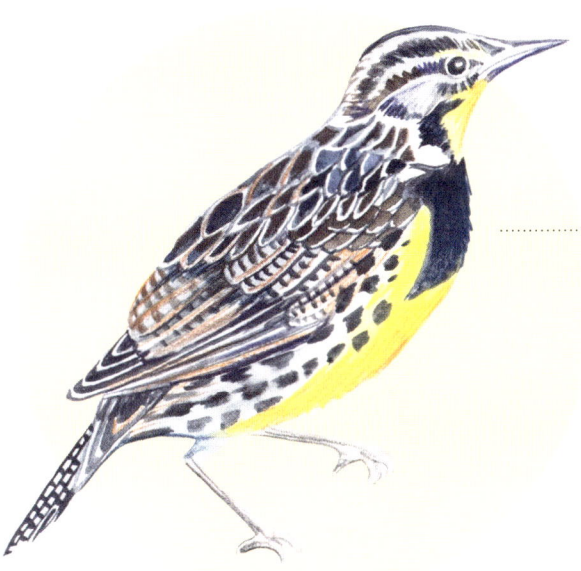

6. & 7.
BLACK-CAPPED CHICKADEE

Poecile atricapillus
Order Passeriformes

The *chickadee-dee-dee* call for which these birds are named is part of a complex system of communication used to indicate threat levels. The more *dee*s in a call, the more serious the threat.

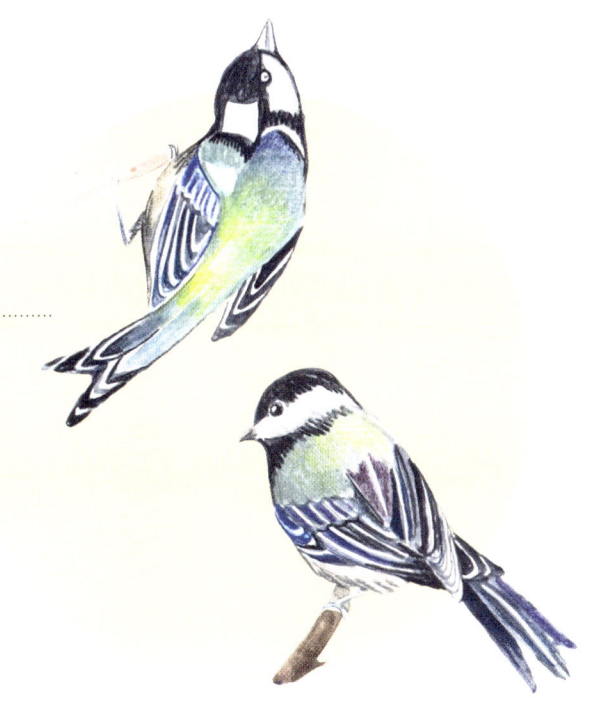

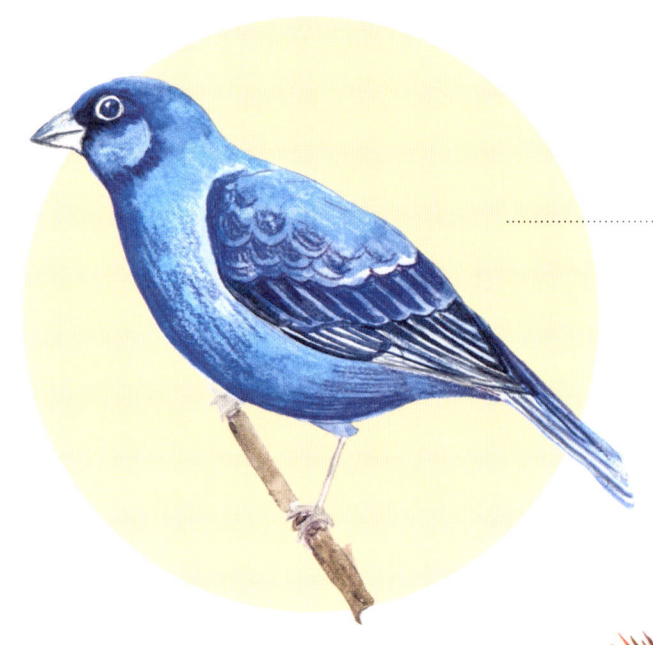

8.
INDIGO BUNTING

Passerina cyanea
Order Passeriformes

This night-migrating species follows the patterns of stars to guide their navigation. They possess an internal clock that helps them continually adjust their orientation to a star, even as the star changes position in the night sky.

9.
NORTHERN CARDINAL

Cardinalis cardinalis
Order Passeriformes

These fiercely territorial birds are so obsessed with defending territory that they have been known to attack their own reflections in window glass and in car mirrors and bumpers.

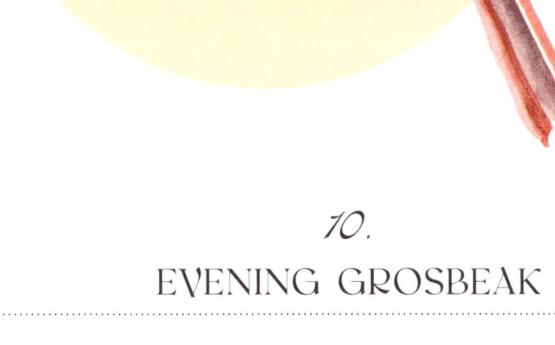

10.
EVENING GROSBEAK

Coccothraustes vespertinus
Order Passeriformes

In summer, this species preys on insects, caterpillars, and worms. In winter, they eat seeds. Their enormous bills enable them to crush seeds that are too large for many other birds to eat.

11.
RUBY-THROATED HUMMINGBIRD

Archilochus colubris
Order Caprimulgiformes

These hummers have good color vision and prefer to feed on the nectar of red and orange flowers. Hummingbird feeders usually feature these colors to attract the tiny birds.

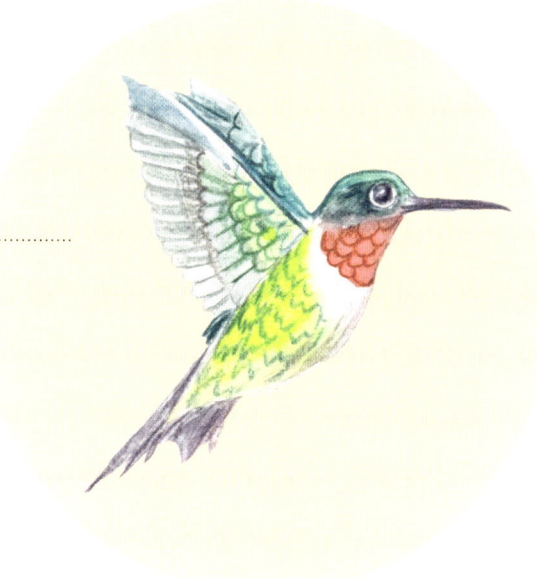

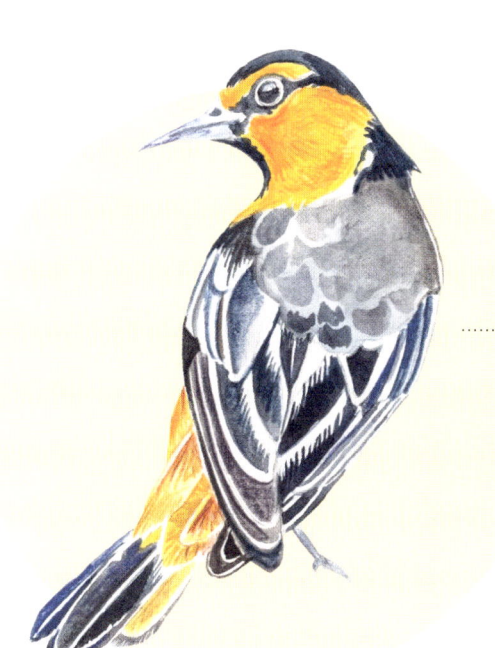

12.
ORCHARD ORIOLE

Icterus spurius
Order Passeriformes

The female of this species is lemon-yellow versus the rusty orange color of the male. These orioles usually nest in groups, with several nests in a single tree.

13.
COMMON GRACKLE

Quiscalus quiscula
Order Passeriformes

In winter, grackles sometimes forage and roost in very large communal flocks with several different species of blackbirds. Some of these flocks can number in the millions!

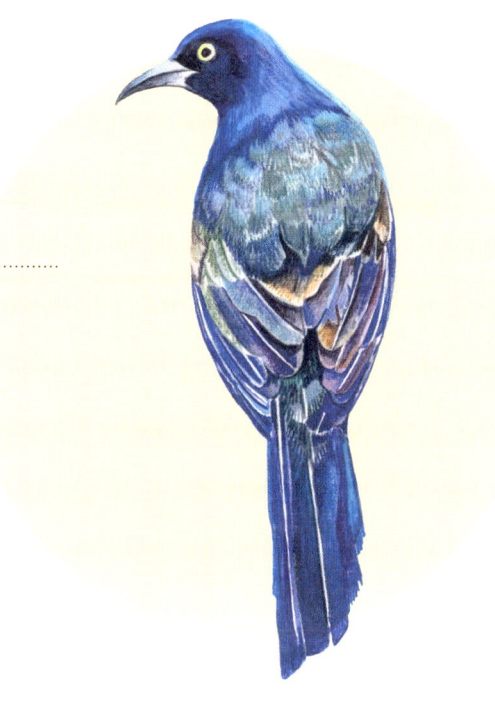

159

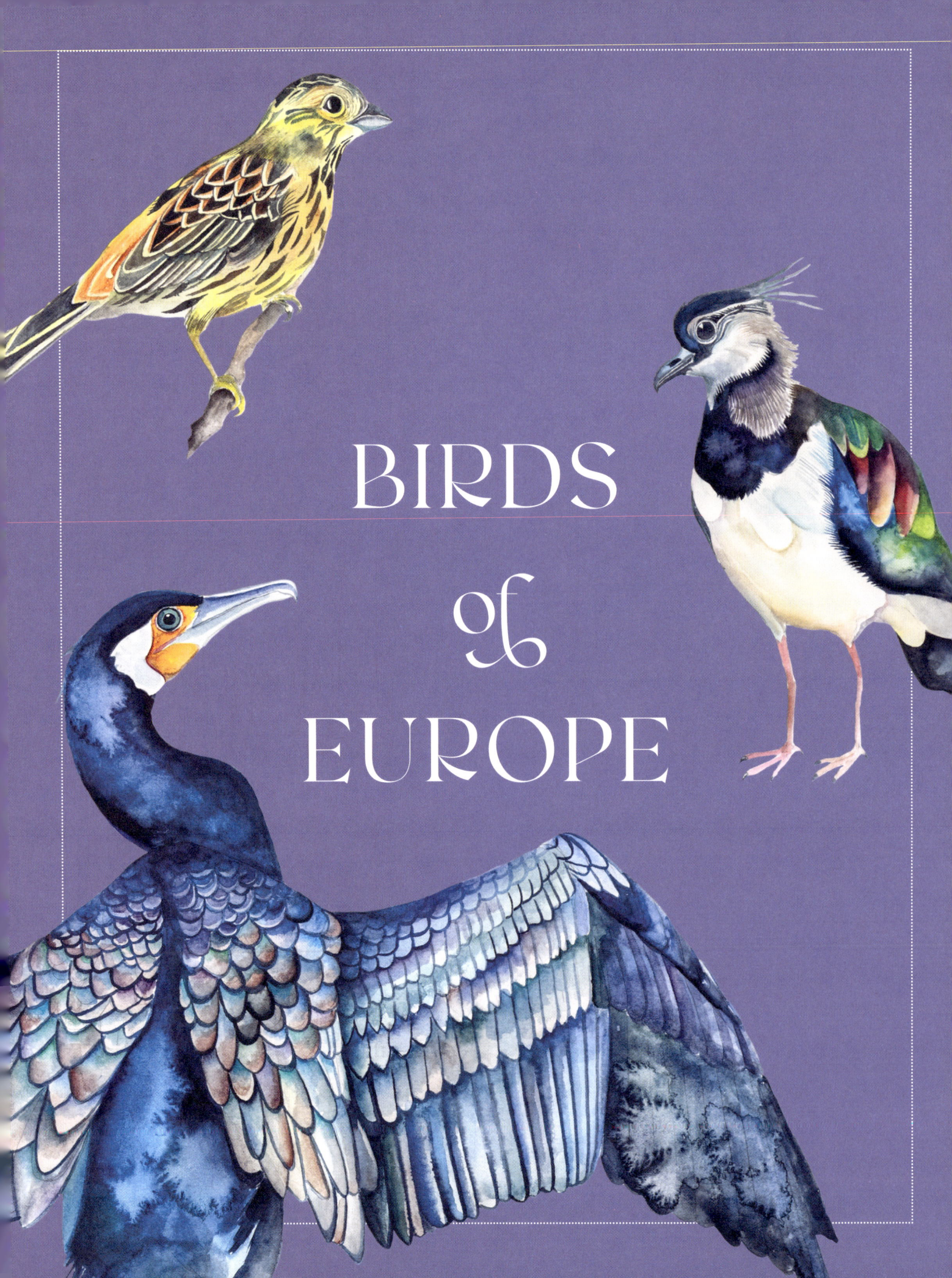

BIRDS
of
EUROPE

From the European Robin, often sighted in British gardens, to the exotic-looking Atlantic Puffin that lives along the northern European coast, the birds of Europe are varied and interesting. Stroll through an English garden and listen for songbirds among the roses. Climb high in the Pyrenees and look for the massive wings of the Eurasian Griffon Vulture flying high above the craggy peaks. Search along a jagged cliff face for a glimpse of the colorful Wallcreeper. Relax near the wetlands of the Danube Delta and watch flocks of pelicans take flight over the water.

BIRDS of BRITISH GARDENS

Among the tangles of foxgloves, hollyhocks, and creeping pink roses are many birds that linger in the frothy grasses and colorful flowers of a classic British garden. You may hear some of these birds before you see them, such as the melodious song of the European Goldfinch (2), so admired for its singing that in nineteenth-century England they were often trapped and caged inside people's homes. You will likely see the national bird of Britain, the European Robin (1). Despite sharing a name and a distinctive red-orange breast with the American Robin, the European Robin is not related to its American counterpart—the American species is in the thrush family, and the European species is an Old World flycatcher.

The European Robin is a ubiquitous character in folklore and mythology throughout Britain and France. In Norse mythology, the god of thunder, Thor, holds the robin as a sacred bird associated with storms. A Christian legend claims that when Jesus was dying on the cross, a robin, then merely a brown bird, perched upon his ear and sang sweet songs to him to distract him from his pain. The blood of Jesus stained the little brown bird's chest, and thereafter it is said that all robins have the red mark of Christ's blood upon their chest. Shakespeare references another famous appearance of the robin in his play *Cymbeline*, wherein the character Arviragus recounts the story from the late sixteenth-century ballad *Babes in the Wood*. In this story, the robin comes across the bodies of deceased children and covers them with leaves and flowers from the forest to honor them.

Another bird that regularly makes an appearance in British gardens is the Eurasian Magpie (3). Magpies are members of the corvid family, highly intelligent birds that also include crows and ravens. In European folklore, an ancient rhyme tells of the superstition that counting the number of magpies one sees foretells certain fates and fortunes for the watcher—"One for sorrow, two for mirth, three for a funeral, and four a birth."

1.	*4.*	*6.*	*9.*	*12.*
EUROPEAN ROBIN	GRAY WAGTAIL	COAL TIT	WHITE WAGTAIL	EURASIAN JAY
2.	*5.*	*7.*	*10.*	*13.*
EUROPEAN GOLDFINCH	EURASIAN GREEN WOODPECKER	EURASIAN BULLFINCH	EURASIAN BLUE TIT	GOLDCREST
3.		*8.*	*11.*	
EURASIAN MAGPIE		EUROPEAN GREENFINCH	COMMON CHAFFINCH	

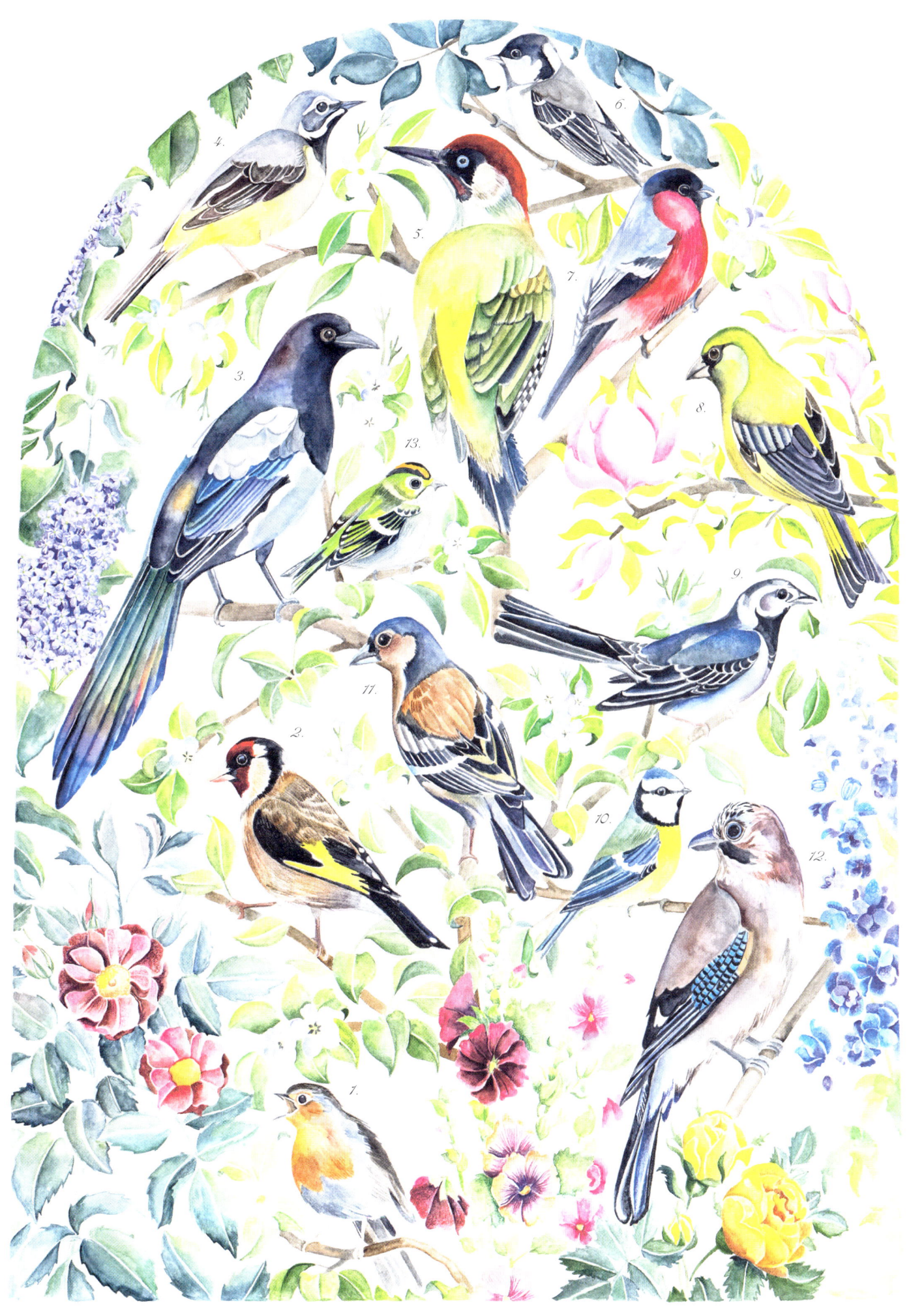

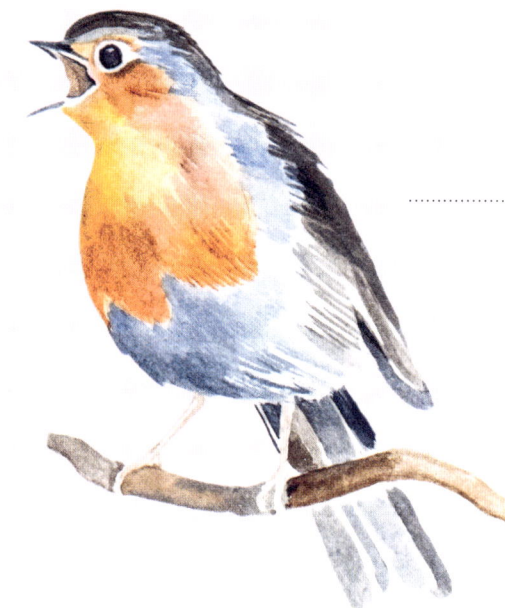

1.
EUROPEAN ROBIN

Erithacus rubecula
Order Passeriformes

These robins are known to follow gardeners
as they work in hopes of snatching
invertebrates stirred up by their digging.

2.
EUROPEAN GOLDFINCH

Carduelis carduelis
Order Passeriformes

The goldfinch appears in many Renaissance
paintings as a symbol of redemption
and the Passion of the Christ.

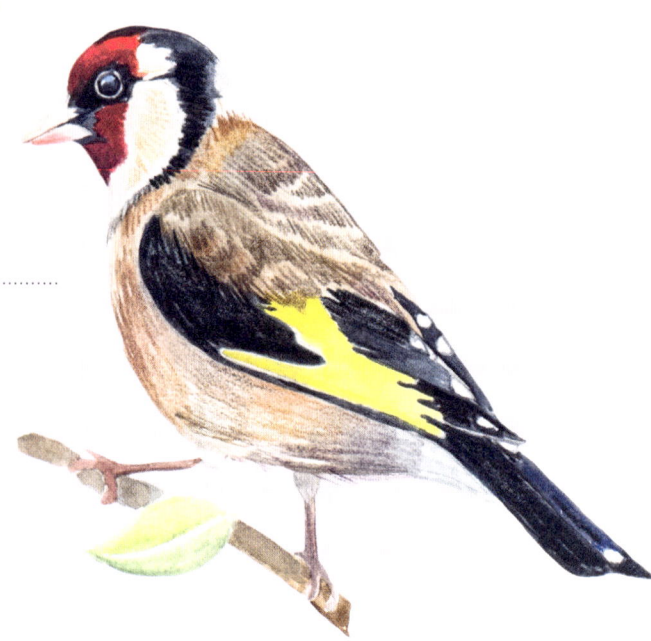

3.
EURASIAN MAGPIE

Pica pica
Order Passeriformes

In medieval France, the call of the magpie was said
to warn of the danger of approaching wolves.

4.
GRAY WAGTAIL

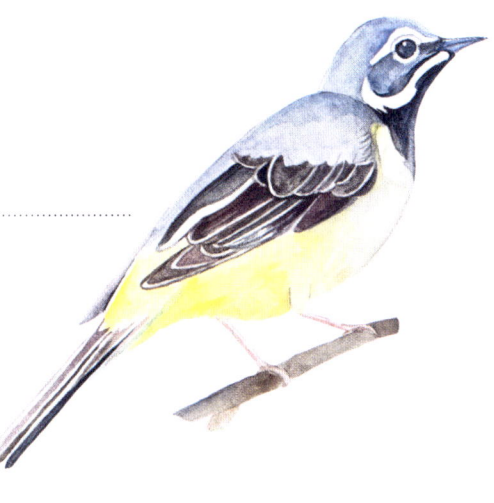

Motacilla cinerea
Order Passeriformes

These birds tend to build their nests near fast-running streams or rivers, often in embankments of stones and roots just above the water.

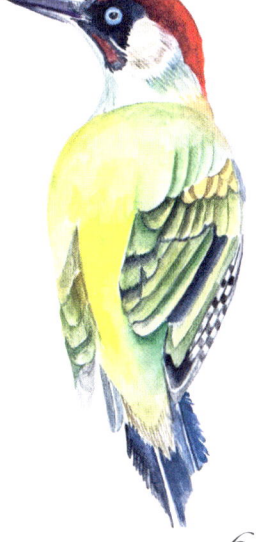

5.
EURASIAN GREEN WOODPECKER

Picus viridis
Order Piciformes

Unlike other woodpecker species, this European bird spends little time drumming on trees. Instead, it tends to feed on ants and insects on the ground.

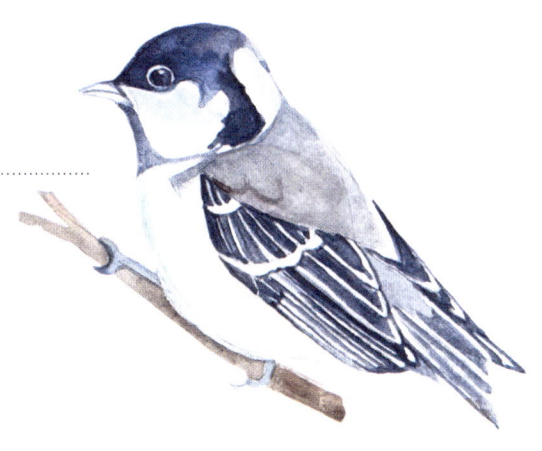

6.
COAL TIT

Periparus ater
Order Passeriformes

When it feels threatened or agitated, the Coal Tit raises a small, hidden crest of feathers on the crown of its head.

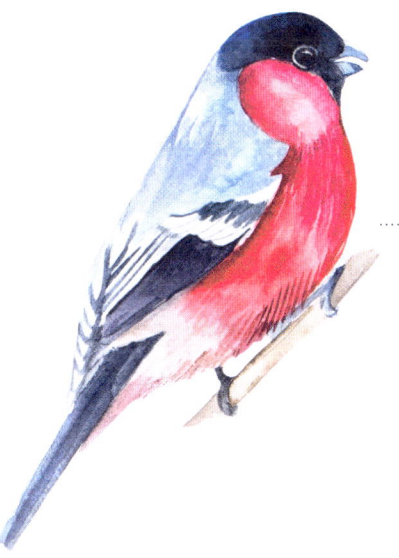

7.
EURASIAN BULLFINCH

Pyrrhula pyrrhula
Order Passeriformes

In nature, these handsome birds sing a low, melodic song; when tamed, they can be taught to sing remarkably complicated melodies.

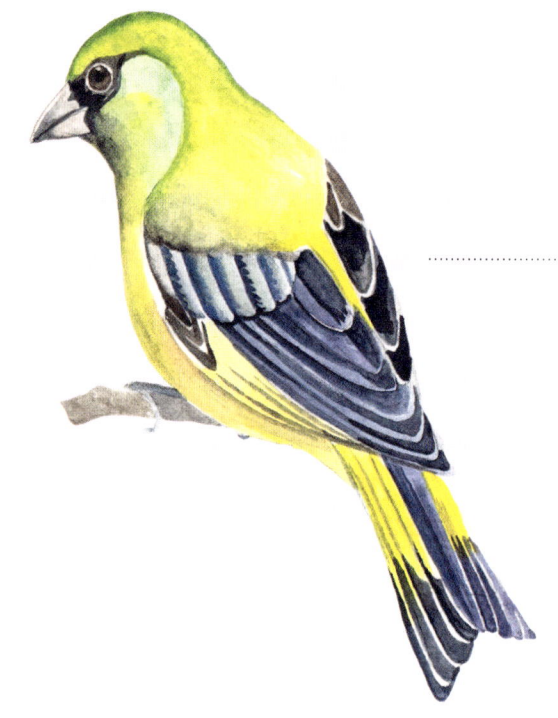

8.
EUROPEAN GREENFINCH

Chloris chloris
Order Passeriformes

These social birds often forage in large groups with other species and are regular and enthusiastic visitors to backyard feeders.

9.
WHITE WAGTAIL

Motacilla alba
Order Passeriformes

Nine subspecies of this ubiquitous species are spread across an immense breeding range, which includes Greenland and Iceland, across Europe and Asia, and in western Alaska.

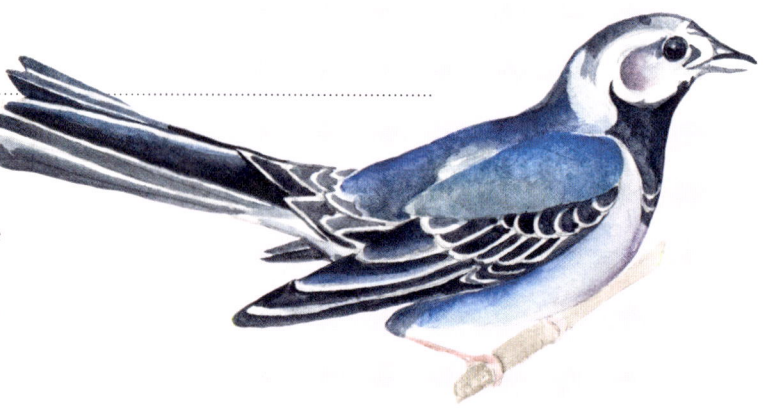

10.
EURASIAN BLUE TIT

Cyanistes caeruleus
Order Passeriformes

These birds are well known for their ability to hang from precarious places while searching for food. They perform elaborate acrobatics, sometimes hanging upside down, to reach hard-to-access insects, fruit, and seeds.

11.
COMMON CHAFFINCH

Fringilla coelebs
Order Passeriformes

Like other finch species, these melodious songbirds were once prized as pets. In Belgium, a popular contest pits two chaffinches against each other to see which will produce the most calls in an hour.

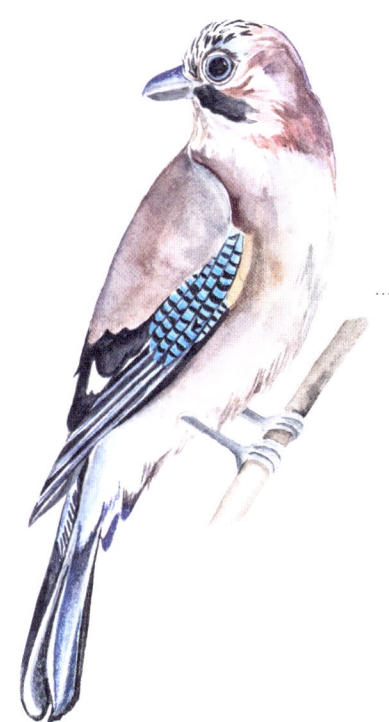

12.
EURASIAN JAY

Garrulus glandarius
Order Passeriformes

This bird's Latin name means "chatty" and "of acorns," referring to its chattering calls and its preferred diet of acorns.

13.
GOLDCREST

Regulus regulus
Order Passeriformes

Despite being one of the smallest birds in Europe, the Goldcrest was awarded the title "king of the birds" in an old folktale, which is reflected in its Latin name (*regulus* comes from the Latin word *rex*, meaning "king").

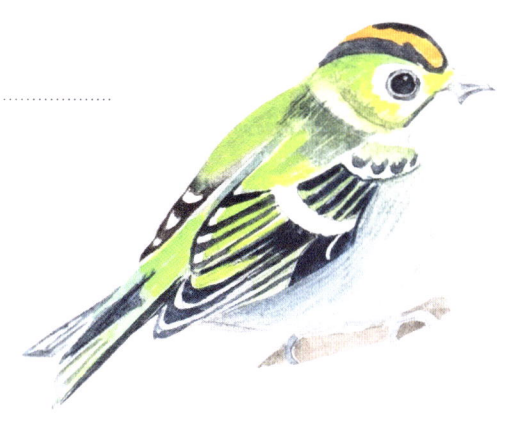

COASTAL BIRDS
of the
ARCTIC

Arctic birds must be uniquely adapted to withstand the unforgiving cold. The Atlantic Puffin (1), for example, is an incredibly distinctive bird with black-and-white waterproof plumage that protects it from the cold water in which it dives to feed. It has a large, unmistakable brightly colored bill that grows larger as it ages. Though the species' range extends throughout the Northern Atlantic, from Norway to Canada, Iceland is the puffin-watching capital and home to more than 60 percent of the world's puffin population. Puffins are featured in various myths, legends, and superstitions. In some cultures, they are believed to predict storms at sea: if puffins fly inland, a storm can be expected to reach land in two or three days. In Cornish legend, King Arthur was reborn as a puffin after his death and was said to reappear in places that held great meaning for him.

Speaking of King Arthur, the Merlin (2) is a small raptor of Nearctic and Palearctic regions. Although this small falcon shares its name with the famous wizard of Arthurian legend, it was not named for him; its name comes from *esmerillon*, the old French name for the species. Merlins are traditionally used in falconry, and medieval falconers called them "lady hawks" because they were often trained to hunt birds for noblewomen.

The Great Cormorant (3) is a very large diving bird that often stands with its wings unfurled, which helps its feathers to dry. In some Nordic countries, the image of a cormorant with its wings unfurled is seen as a symbol of protection—but take caution, because in Ireland, seeing a cormorant perched atop a church steeple is said to bring bad luck.

The Rock Ptarmigan (8) changes color with the seasons to blend in with its environment. In the coldest part of the year, its plumage is snow-white, but as summer approaches, it changes to tones of deep brown. In spring, these birds are known to disguise their still-white feathers by covering themselves in dirt until they have safely molted into their summer browns. This is the official bird of Nunavut, the largest and northernmost territory of Canada, another region with plummeting winter temperatures.

1. ATLANTIC PUFFIN	*3.* GREAT CORMORANT	*5.* REDPOLL	*7.* WHIMBREL
2. MERLIN	*4.* NORTHERN WHEATEAR	*6.* SNOW BUNTING	*8.* ROCK PTARMIGAN

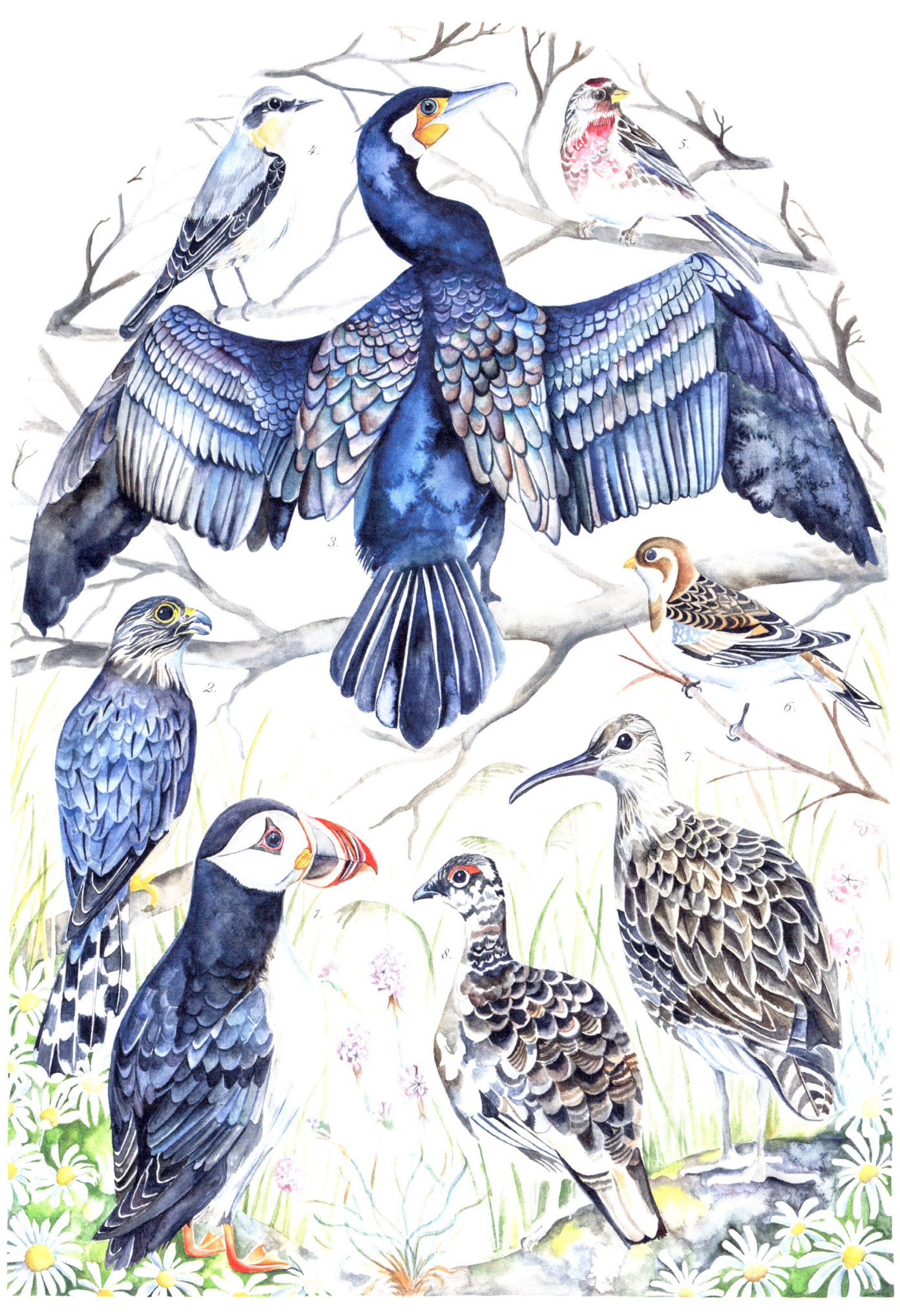

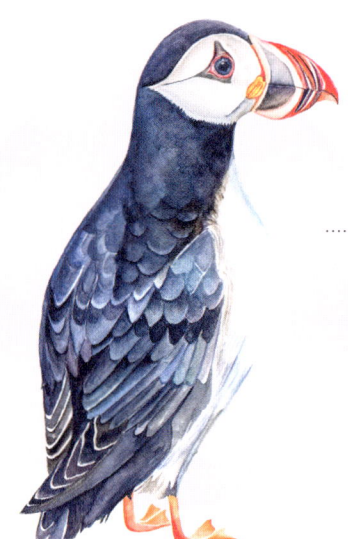

1.
ATLANTIC PUFFIN

Fratercula arctica
Order Charadriiformes

Puffins are monogamous, and once paired, they mate for life. The pair may stay together for more than twenty years. They maintain their bond by rubbing their bills together.

2.
MERLIN

Falco columbarius
Order Falconiformes

Merlins don't build their own nests. Instead, they forcibly take over the nests of other raptors, crows, and magpies.

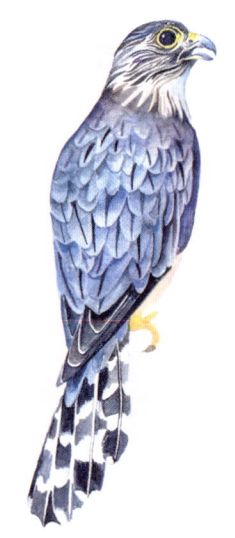

3.
GREAT CORMORANT

Phalacrocorax carbo
Order Suliformes

People have used cormorants to help them fish for centuries. In sixteenth-century Europe, fishermen would fit rings on the birds' necks to prevent them from swallowing the fish they caught.

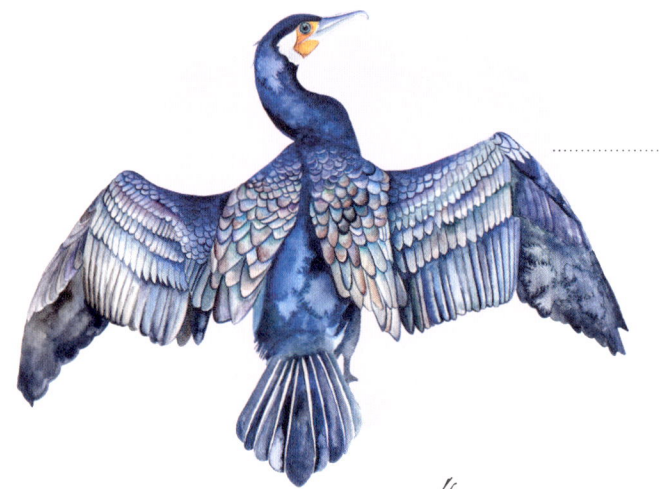

4.
NORTHERN WHEATEAR

Oenanthe oenanthe
Order Passeriformes

In the eighteenth and nineteenth centuries, these birds were eaten as a delicacy. Some small songbirds, such as the Orlotan Bunting, are still being consumed in Europe, especially France, although it is illegal to capture, kill, or sell them.

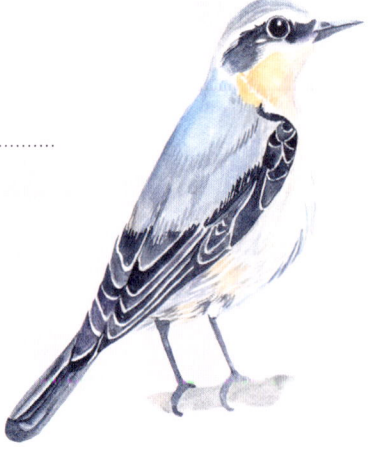

5.
REDPOLL

Acanthis flammea
Order Passeriformes

During the coldest nights of winter, Redpolls burrow beneath the snow to stay warm. They are known to dig tunnels that can be more than a foot long.

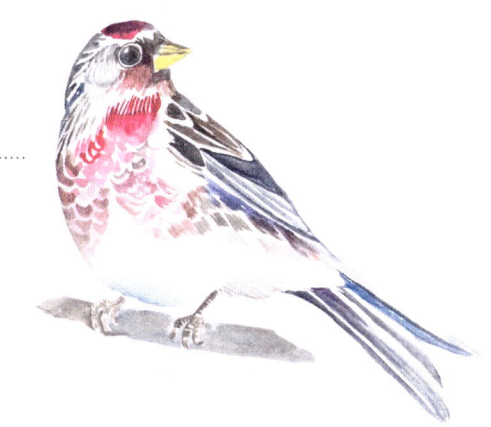

6.
SNOW BUNTING

Plectrophenax nivalis
Order Passeriformes

In winter, these birds' feathers are brownish, which helps hide them in fields and along the water's edge. During summer breeding season in the high arctic, males rub their bellies and heads on the snow to wear down the brown feather tips and reveal the snow-white feathers underneath.

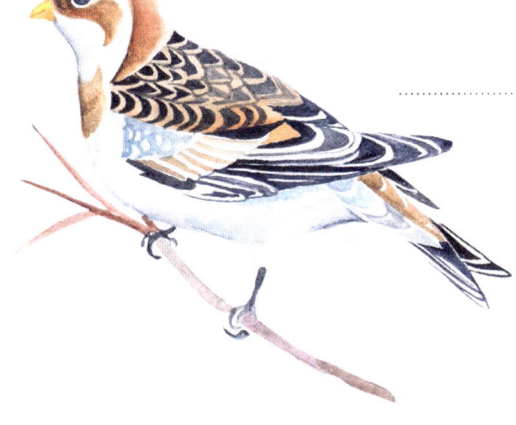

7.
WHIMBREL

Numenius phaeopus
Order Charadriiformes

The genus name for this bird is Greek for "new moon," which is a reference to the crescent moon shape of its bill.

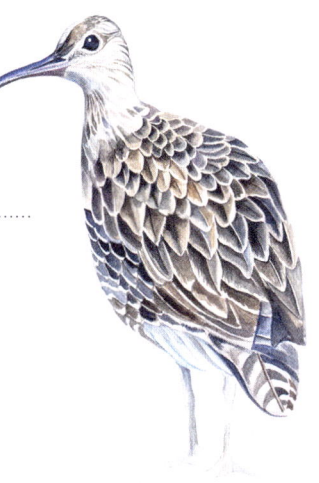

8.
ROCK PTARMIGAN

Lagopus muta
Order Galliformes

In Icelandic culture, ptarmigans are hunted and eaten as a traditional festive meal on Christmas Eve. However, their hunting season is strictly controlled, and they can be hunted only for personal consumption; all trade and sales of the bird for meat are forbidden.

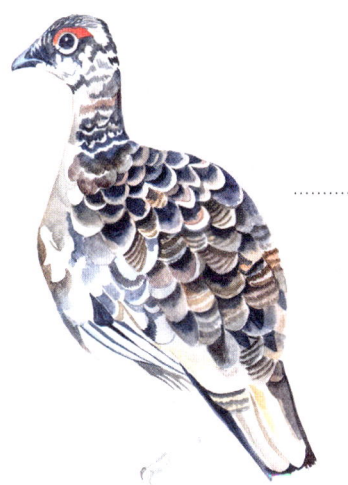

FOREST BIRDS of NORTHERN EUROPE

The Western Capercaillie (2), the largest member of the grouse family, is a prized game bird once ubiquitous in Scotland and Germany and today found mostly in Scandinavia. Hunting the bird is now illegal or restricted because of declining populations. Its name comes from the Scottish Gaelic *capal coille*, which means "horse of the forest." This bird can be as large as a turkey! It keeps to mossy forests, where it feeds from the shrubs and berries of spruce and pine. In Finnish folktales, it was associated with witches and magic-makers, and it was believed that various fairies and enchanters could take the form of a capercaillie.

Another bird that often appears in mythology and literature is the European Turtle-Dove (5), mentioned by Shakespeare and referenced as a companion to the Greek goddess Demeter. The goddess of the harvest, associated with birth and fertility, is often depicted with a dove. In mythology, her daughter, Persephone, was stolen by Hades, god of the underworld.

German composer Ludwig van Beethoven was known for his great love of the outdoors and proclivity for taking long nature walks. During these meditative walks, he was said to have gathered inspiration for his symphonies from several beautifully singing birds. It has been suggested that the song of the Yellowhammer (1) was the inspiration for the first four notes of his Fifth Symphony. Eurasian Blackbirds (6) also would have fluttered across his path and are believed to have "struck a chord" with the wandering composer.

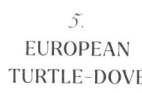

1.
YELLOWHAMMER

2.
WESTERN CAPERCAILLIE

3.
EUROPEAN STARLING

4.
BLUETHROAT

5.
EUROPEAN
TURTLE-DOVE

6.
EURASIAN BLACKBIRD

7.
EURASIAN GOSHAWK

8.
DUNNOCK

9.
PINE GROSBEAK

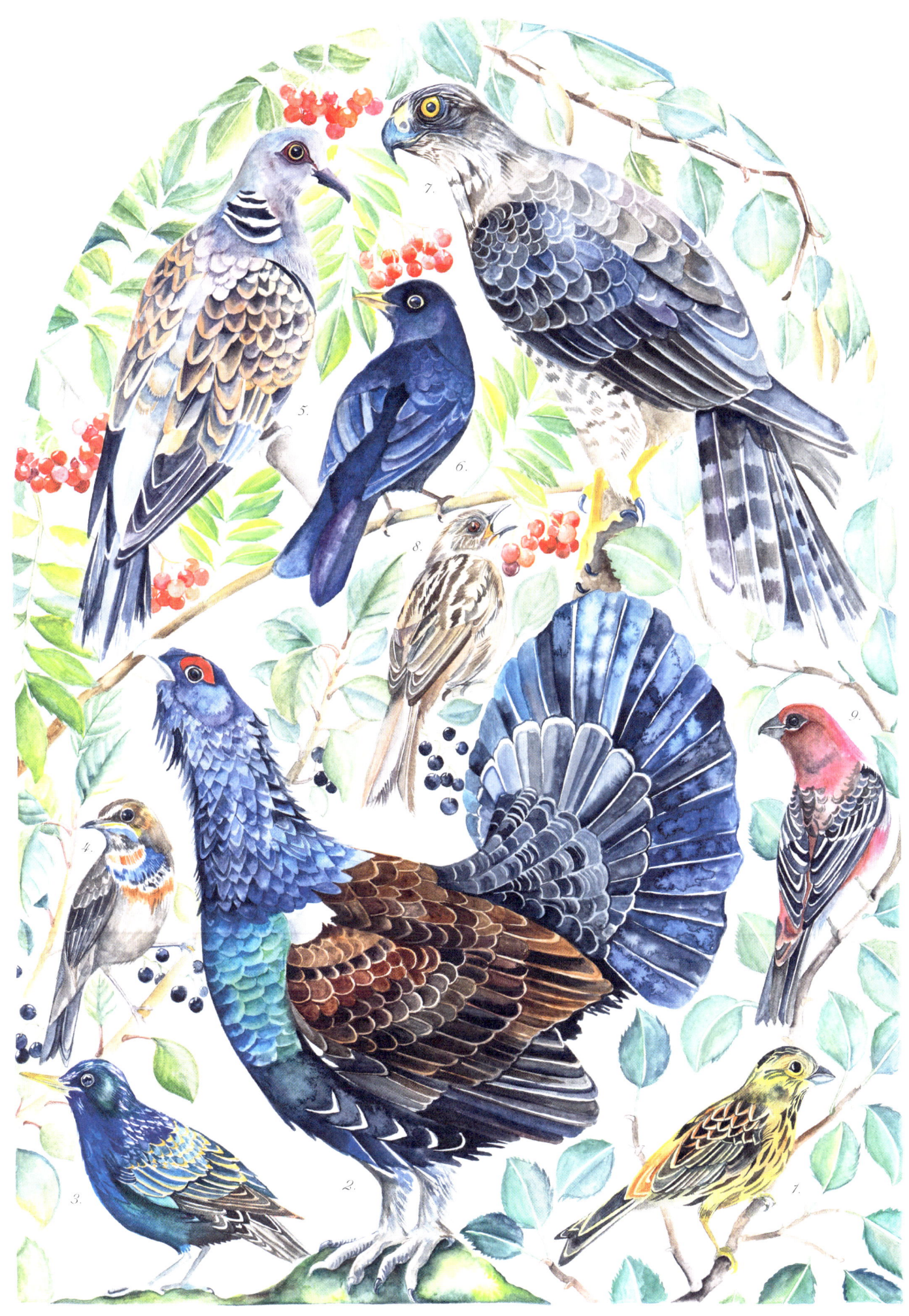

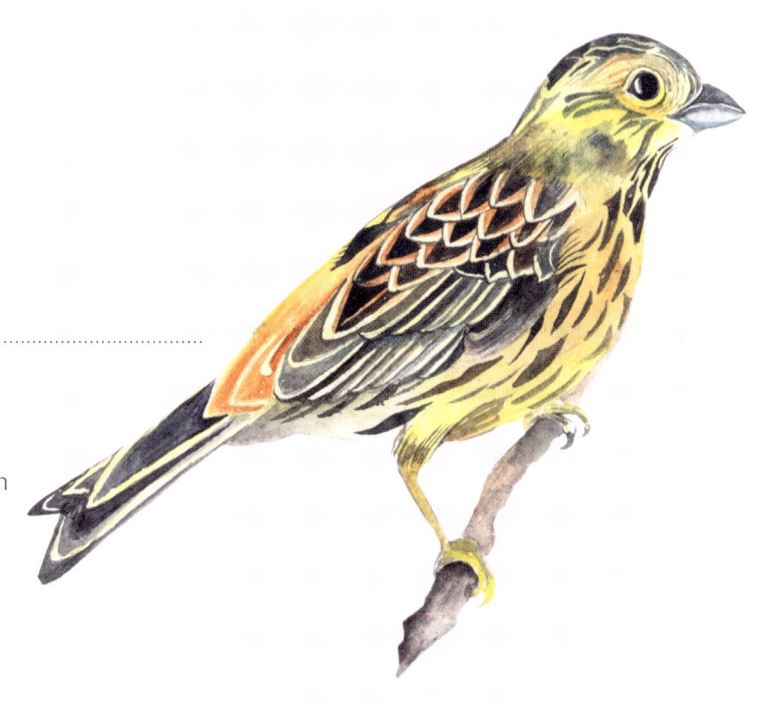

1.
YELLOWHAMMER

Emberiza citrinella
Order Passeriformes

An old legend tells of the Yellowhammer's association with the devil; its tongue was said to possess a drop of satanic blood, and the pattern on its eggs was said to carry a secret message. These beliefs sometimes led to persecution of the bird.

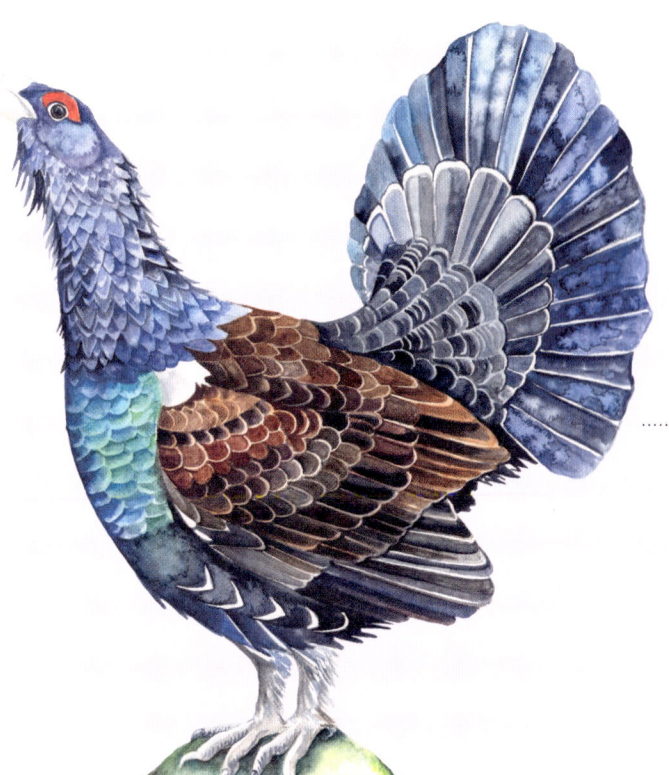

2.
WESTERN CAPERCAILLIE

Tetrao urogallus
Order Galliformes

Male capercaillies perform elaborate mating dances to attract females. Competition among males can become quite aggressive, sometimes even resulting in birds killing one another as they compete for a female's attention.

3.
EUROPEAN STARLING

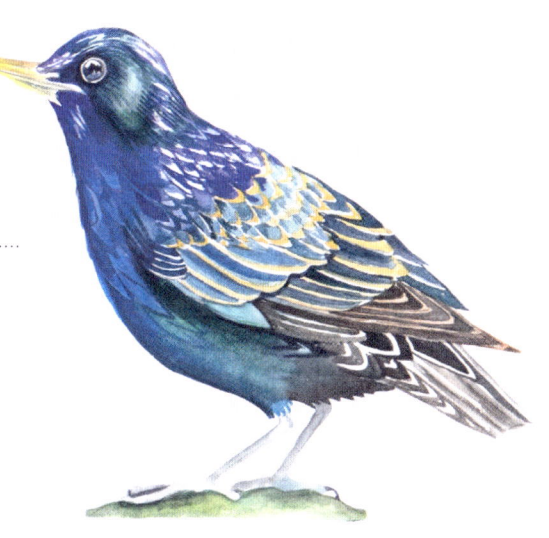

Sturnus vulgaris
Order Passeriformes

These birds can mimic other birds' songs. A starling
can learn the calls of up to twenty different species.

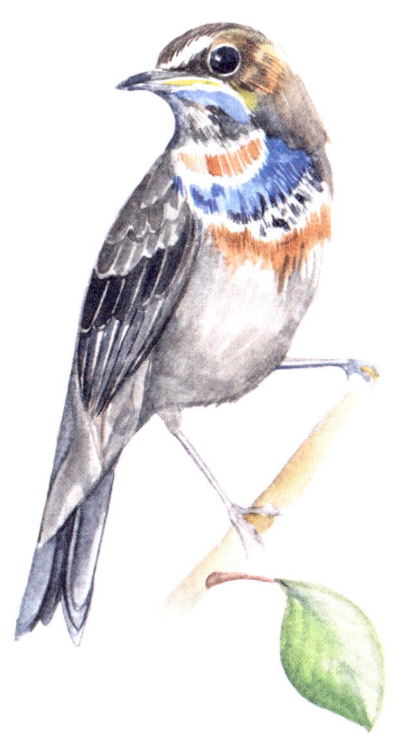

4.
BLUETHROAT

Luscinia svecica
Order Passeriformes

These striking birds are secretive and difficult
to spot, as they generally stay hidden in trees,
but you may hear their song. If you do see
one, it is likely to be flying overhead.

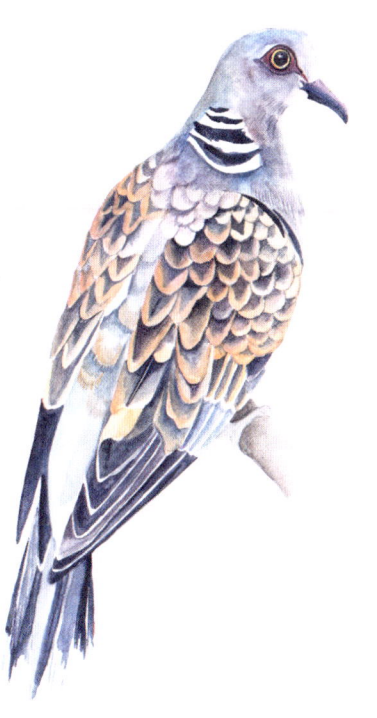

5.
EUROPEAN TURTLE-DOVE

Streptopelia turtur
Order Columbiformes

This small dove appears in the New Testament
as a symbolic herald of the coming spring. Today,
populations of this species are in decline as a result
of habitat loss, climate change, and hunting.

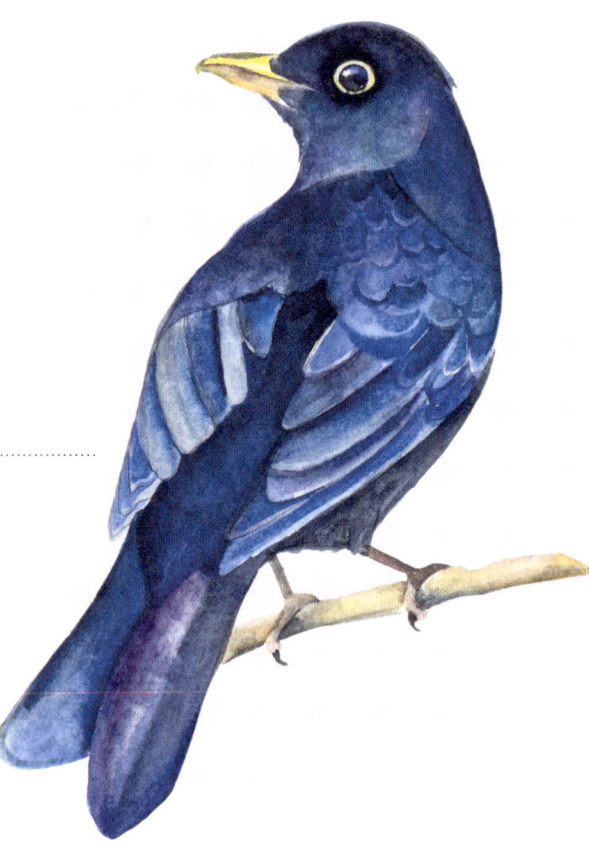

6.
EURASIAN BLACKBIRD

Turdus merula
Order Passeriformes

Some cultures view these glossy, elegant birds as symbols of bad luck or danger, while others consider them harbingers of good luck and prosperity.

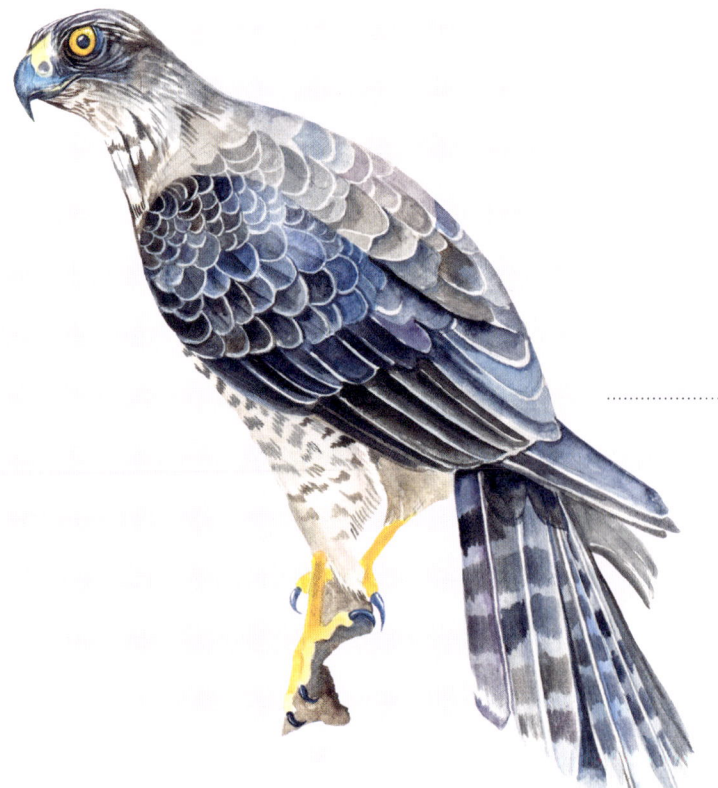

7.
EURASIAN GOSHAWK

Astur gentilis
Order Accipitriformes

Known as tenacious hunters, these birds have a long history in falconry. In the wild, they are flexible hunters and will take prey as large as grouse, squirrels, and hares. They also eat other birds.

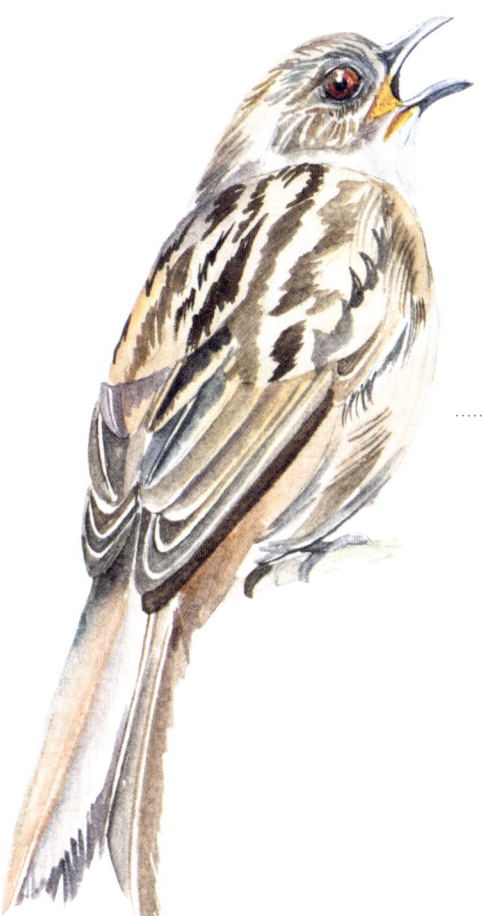

DUNNOCK

Prunella modularis
Order Passeriformes

This robin-sized bird was successfully introduced to New Zealand in the nineteenth century. They are "flexible" breeders—sometimes monogamous, but some populations mate freely with several partners.

9.

PINE GROSBEAK

Pinicola enucleator
Order Passeriformes

These birds are often remarkably tame and unafraid of humans. They eat mostly plants and feed their nestlings a regurgitated paste of insects and vegetable matter that they store in pouches inside their mouths.

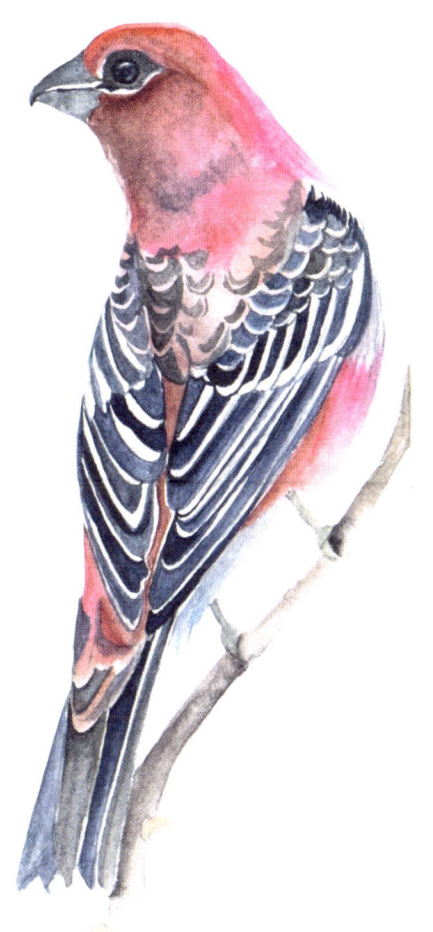

BIRDS of the PYRENEES

The Pyrenees, a chain of precipitous, craggy mountains situated along the border of France and Spain, are some of the tallest peaks in all of Europe. Boasting brilliantly turquoise lakes and a network of cascading waterfalls, and trickling with icy mountain streams, the region is visited by many who hike and take in the striking views. While struggling up the steep trails, mountaineers can spot a variety of alpine birds.

Scan the rocky ledges and you may catch the flutter of red wings that reveals a Wallcreeper (2), a gorgeous little rock climber that breeds high in the mountains. It is a member of the family Tichodromidae, a name derived from the ancient Greek for "which runs on walls," owing to the bird's unique practice of skittering along the sides of jagged and vertiginous mountain ledges in search of insect prey hidden in the crevices. Though Wallcreepers spend most of the year high in the mountains, in the coldest part of winter they retreat to lower altitudes, where they may be seen scaling buildings instead of mountainsides.

The Yellow-billed Chough (4) is a member of the corvid family, along with crows and ravens. These noisy and conspicuous birds can become relatively tame around ski resorts and towns, where they scavenge for food. It is one of only two species in its genus, the other being the Red-billed Chough. The two look remarkably similar, apart from bill color and shape.

One of the largest and most intimidating species of the Pyrenees is the massive Eurasian Griffon (5), an Old World vulture that perches high on cliffs and soars on thermal updrafts over valleys and mountainsides in search of carrion, its preferred food. Vultures, unlike other birds of prey, are voracious and indiscriminate in what they eat. They feed primarily on the flesh of dead animals and do little hunting of their own. This tendency is responsible for the vulture's reputation as a greedy, gluttonous scavenger, and the practice of referring to humans who have these qualities as "vultures" dates back as far as Ancient Rome.

Another predatory bird with a much less impressive size, the Red-backed Shrike (6) is a handsome, compact bird of prey that does its own hunting and uses a somewhat draconian method to finish its kills: it impales insects and small animals on thorns and sharp branches before eating them, or it stores the impaled creatures to eat later. Its Latin genus name, *Lanius*, is derived from the Latin for "butcher," and some refer to these birds as "butcher birds" due to their brutal hunting tactics.

1.
ALPINE ACCENTOR

2.
WALLCREEPER

3.
EUROPEAN STONECHAT

4.
YELLOW-BILLED CHOUGH

5.
EURASIAN GRIFFON

6.
RED-BACKED SHRIKE

7.
CITRIL FINCH

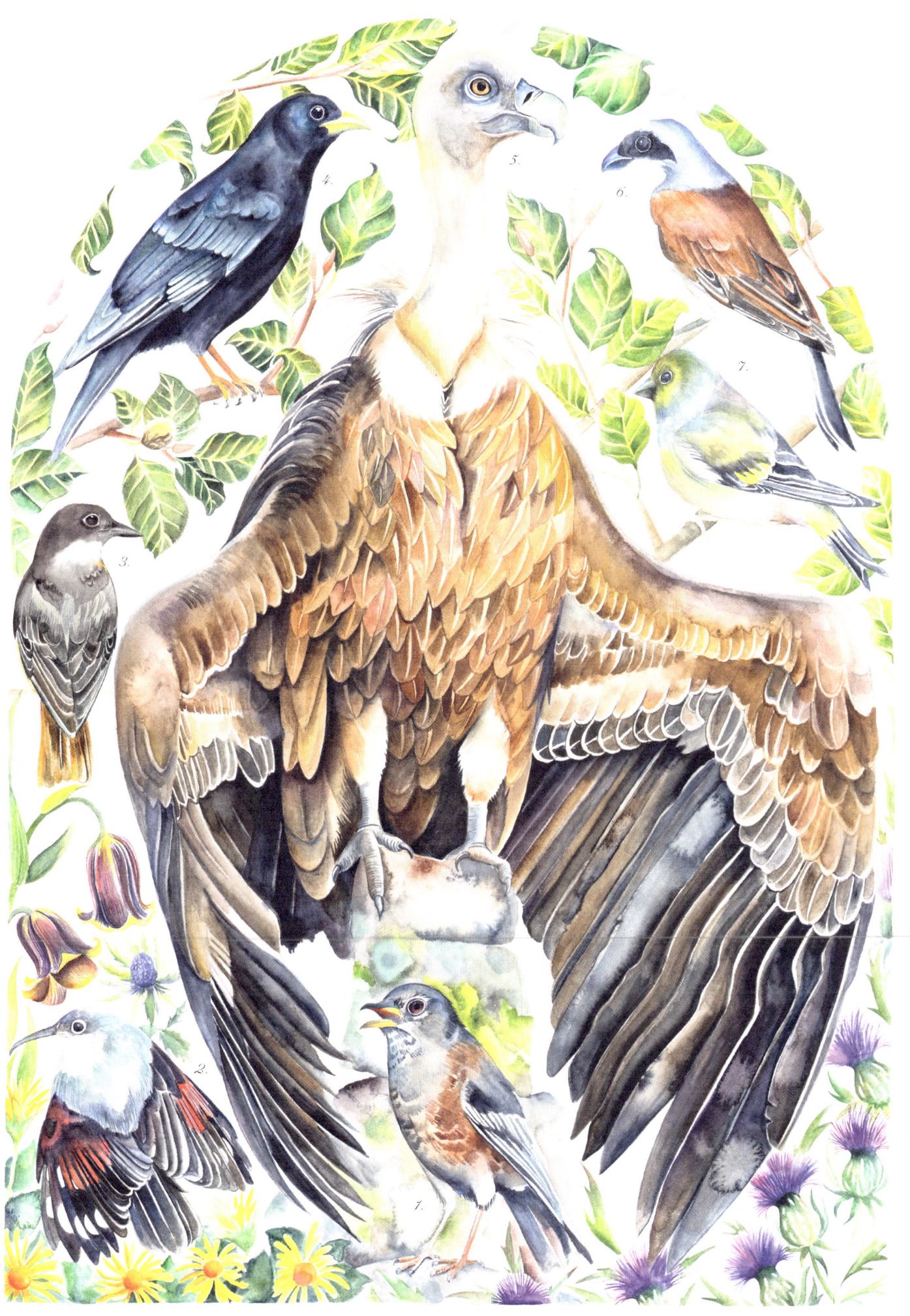

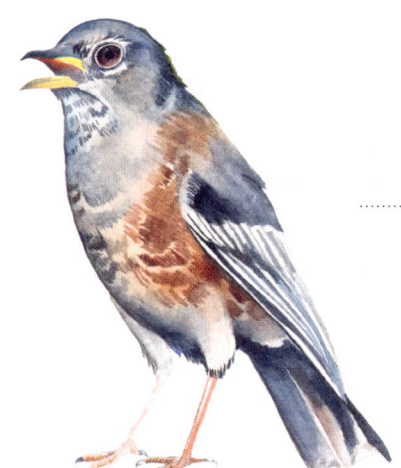

1.
ALPINE ACCENTOR

Prunella collaris
Order Passeriformes

After diving off high rocks and ledges, these birds often launch impressively in acrobatic flight.

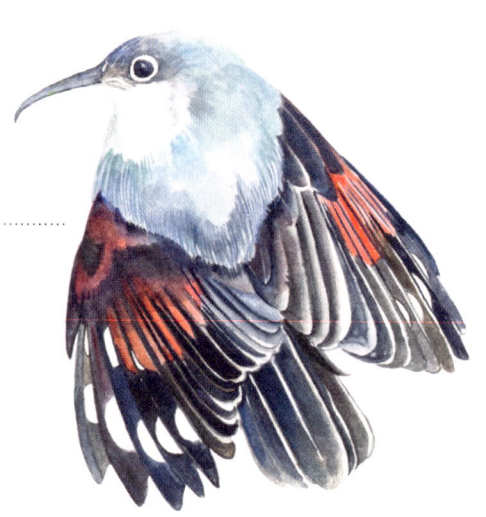

2.
WALLCREEPER

Tichodroma muraria
Order Passeriformes

Some mountaineers call Wallcreepers "butterfly birds" because their darting flight pattern and colorful wings are reminiscent of butterflies.

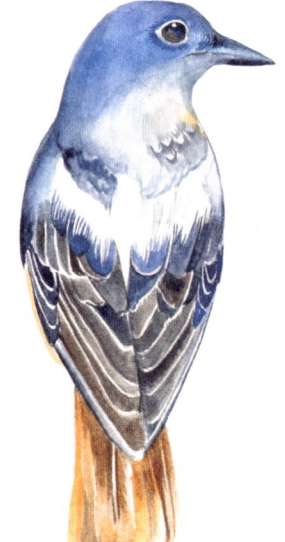

3.
EUROPEAN STONECHAT

Saxicola rubicola
Order Passeriformes

The genus name of this species translates roughly to "rock dwelling." Its common name comes from its call, which sounds like two stones being knocked together.

4.
YELLOW-BILLED CHOUGH

Pyrrhocorax graculus
Order Passeriformes

These noisy and social birds are known for their acrobatic flight— they dive and twirl while soaring in the updrafts along cliff faces.

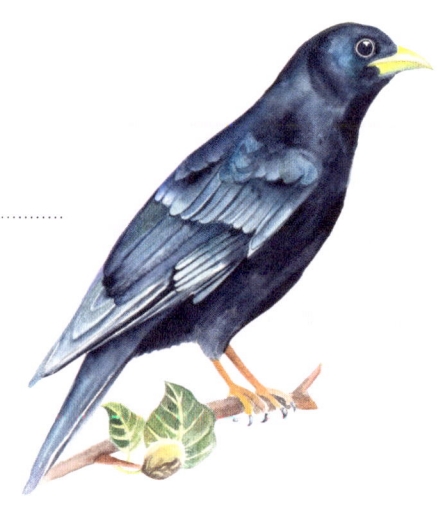

5.
EURASIAN GRIFFON

Gyps fulvus
Order Accipitriformes

After building a nest on a cliff ledge or in a small cave, the female griffon lays a single egg. Both parents take turns incubating the egg and searching for carrion.

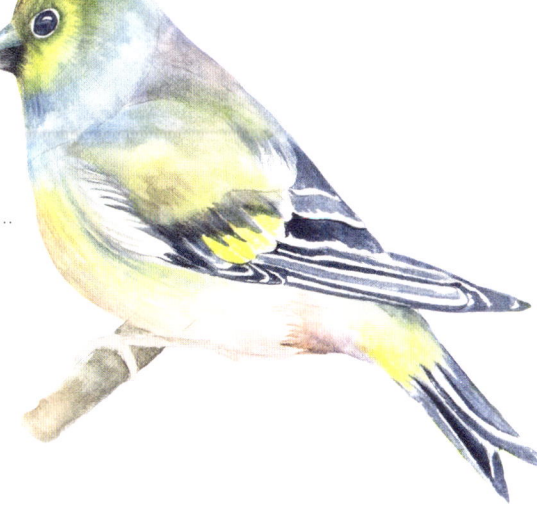

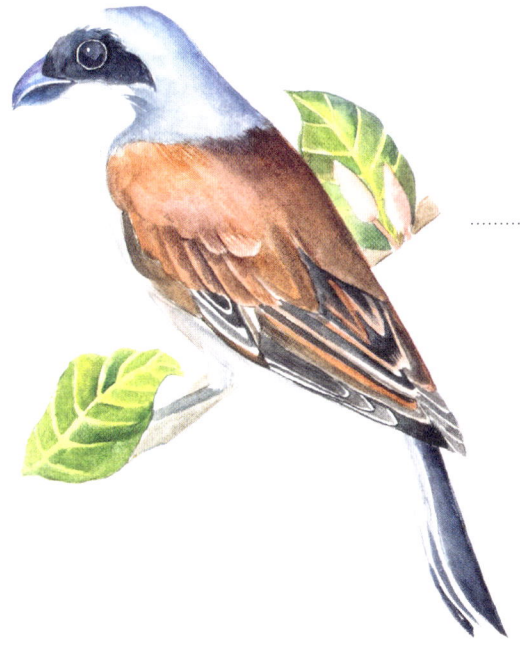

6.
RED-BACKED SHRIKE

Lanius collurio
Order Passeriformes

The impaled prey of this shrike may be stored on thorns, broken twigs, or barbed wire in caches, or larders, that are widely scattered throughout its territory.

7.
CITRIL FINCH

Carduelis citrinella
Order Passeriformes

This species is related to the European Goldfinch, though the two look nothing alike. Their twittering songs, however, are very similar.

BIRDS of the DANUBE DELTA

✳

This large river delta lies mostly within Romania and is home to many impressive waterbirds and migratory visitors, making it a popular destination for birdwatchers from around the world. The waters of the Danube Delta, a labyrinthine system of rivers and lakes that flows into the Black Sea, provide ample habitat for birds to assume residence.

The largest bird featured here, the Great White Pelican (1), has a huge bill and a large throat pouch. Since medieval times, pelicans have been symbols of self-sacrifice. Pelican mothers were said to wound their own breasts to feed their young with their blood, but this is a misconception. A mother pelican collects as many fish as possible in her throat pouch in anticipation of feeding her young. Back at her nest, she presses her beak into her chest to push the fish forward, where her nestlings can grab it from her bill. Sometimes the uncoordinated, over-enthused babies peck at her chest in anticipation of the meal, which causes injury. The pelican's association with sacrifice and piety was not easily shaken, however, and they continued to appear in works of literature and drama as symbols of sacrifice for many years, including references in several of Shakespeare's plays.

Also mentioned by Shakespeare is the heron (which he referred to as a "handsaw"). Two heron species are found in the Danube Delta: the Squacco Heron (3) and the Purple Heron (4). The word *heron* is derived from an old German word for the bird, *heiger*, which means "to utter sharp cries," and certainly applies to these noisy birds. Although many species of heron are found throughout the world, they all share certain characteristics, particularly in flight—their long, elegant legs are extended behind their bodies and their graceful, snakelike necks form an S-shape. In medieval Europe, knights were known to hunt herons for food, giving the bird pride of place at the banquet table after a battle.

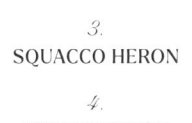

1.
GREAT WHITE PELICAN

2.
NORTHERN LAPWING

3.
SQUACCO HERON

4.
PURPLE HERON

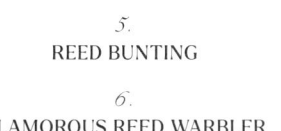

5.
REED BUNTING

6.
CLAMOROUS REED WARBLER

7.
COMMON KINGFISHER

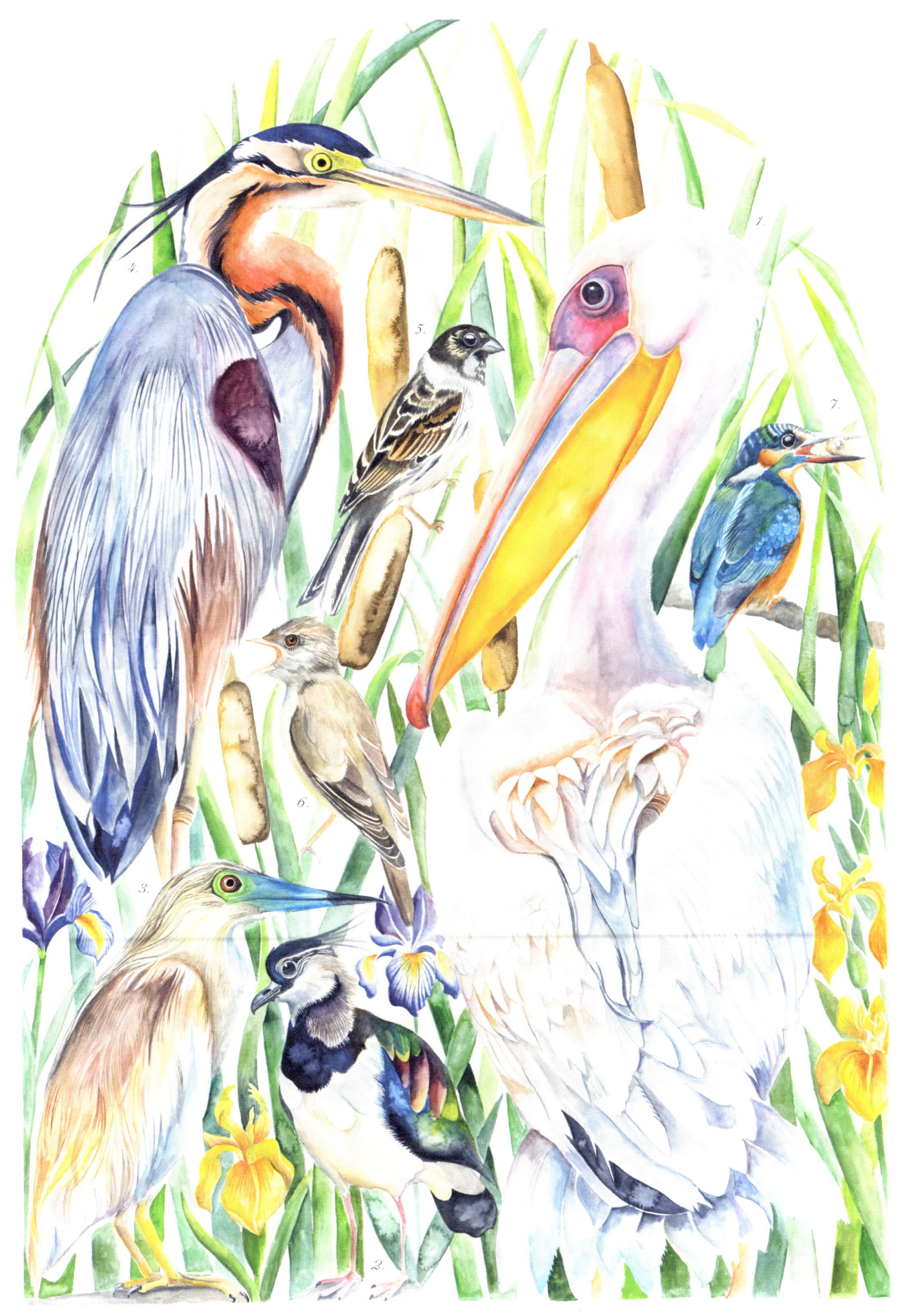

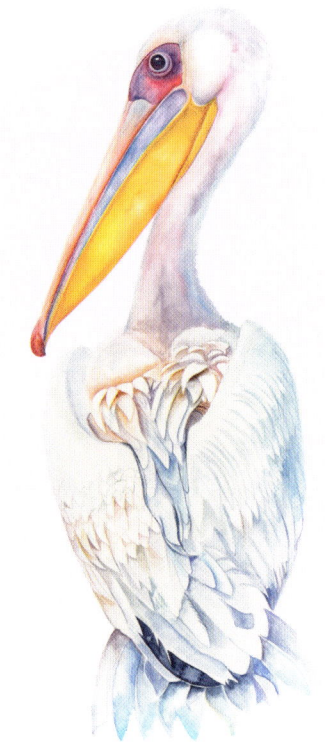

1.
GREAT WHITE PELICAN

Pelecanus onocrotalus
Order Pelecaniformes

The Great White Pelican is the second largest species of pelican in the world, after the Dalmatian Pelican.

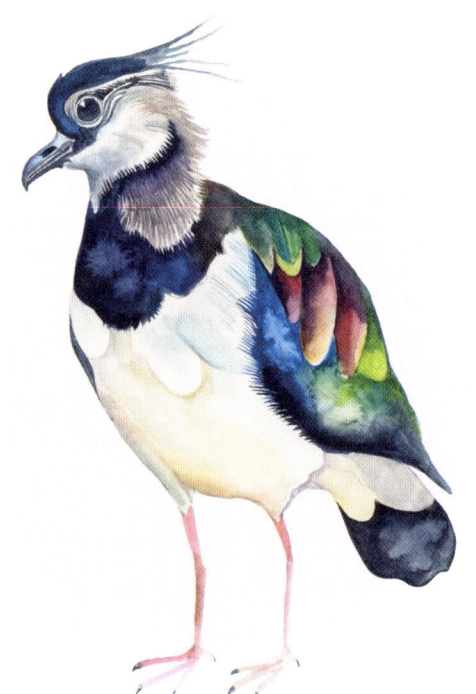

2.
NORTHERN LAPWING

Vanellus vanellus
Order Charadriiformes

The speckled eggs of this handsome wading bird were considered a delicacy among the aristocracy in Victorian England, thanks to Queen Victoria's fondness for them.

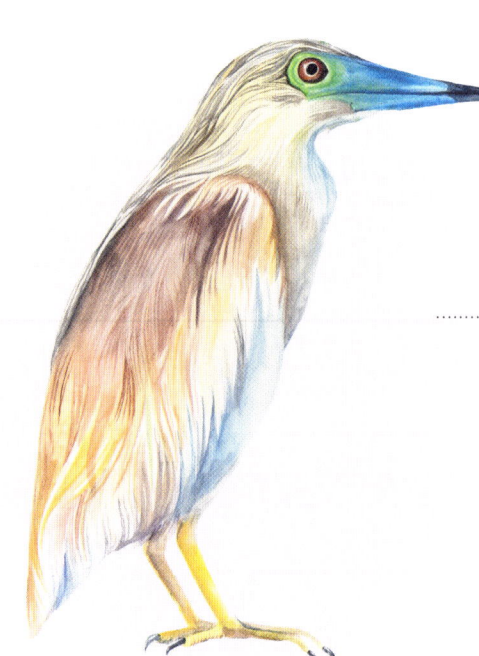

3.
SQUACCO HERON

Ardeola ralloides
Order Pelecaniformes

This small, stout heron is found in freshwater marshes, lakes, and ponds. In flight, this brown bird looks white due to the color of its wings.

4.
PURPLE HERON

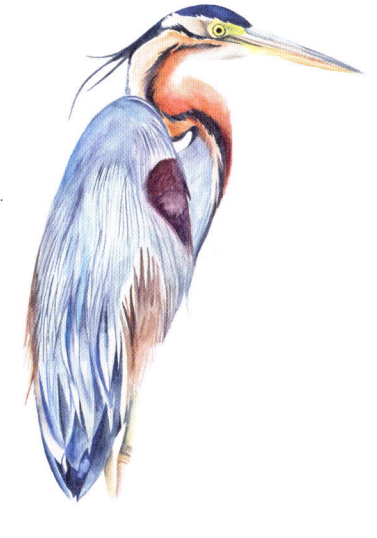

Ardea purpurea
Order Pelecaniformes

These herons are solitary hunters but nest in colonies. They feed in shallow water, often waiting motionless as they stalk prey. They build bulky stick nests in reed beds or trees close to water.

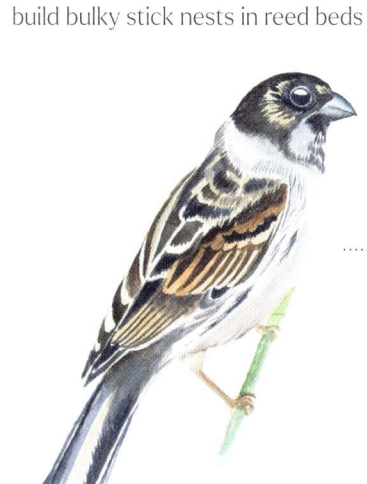

5.
REED BUNTING

Emberiza schoeniclus
Order Passeriformes

Although these sparrow-sized birds are often found in reedy wetland habitats, they also live in grasslands, farmland, and meadows, where they eat seeds and insects.

6.
CLAMOROUS REED WARBLER

Acrocephalus stentoreus
Order Passeriformes

This species' song is a loud, slow, chattering, scratchy *karra-karra-kereet-kereet*, repeated three or four times. Some individuals also mimic other birds.

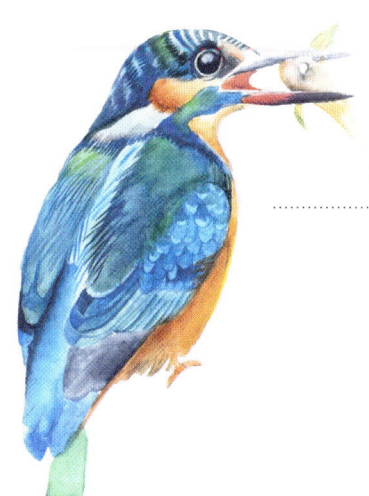

7.
COMMON KINGFISHER

Alcedo atthis
Order Coraciiformes

When it spots a fish, the kingfisher dives headfirst into the water or sometimes hovers before plunge-diving. When it returns to its perch, it may beat the fish's head on a branch to stun it before swallowing it.

BIRD RELATION- SHIPS

Although we may all know and recognize a peacock, a common pheasant, and a domestic chicken, some of their lesser known relatives may be new to us. Eggs are eaten throughout the world, yet it is not widely known that the chicken we recognize today is in fact a domesticated relative of a wild jungle fowl. Here, we will look at a variety of birds and the relationships between them. This includes some of the world's most formidable birds of prey, some of the most beautiful and mysterious nocturnal owls, and a few birds that are not capable of flight for various evolutionary reasons. Many of these birds appear in myths and legends throughout the world and share symbolic significance with cultures separated by thousands of miles. We'll visit just a few of the fascinating avian groups and discuss their relationships and importance throughout the world.

DUCKS

Ducks gracefully glide across the surface of the water or waddle awkwardly through the reeds along the banks. Wild ducks are a favorite of hunters, and most varieties of domestic ducks, which are descended from wild Mallards, are today bred mainly for meat and eggs. Exotic and beautiful ducks have also been fashionable additions to ponds and lakes of nobility. This trend possibly accounts for the presence of various exotic species of waterfowl outside of their original range.

Ducks have cultural significance the world over and pop up in many well-known idioms and sayings: "lucky duck," "sitting duck," "odd duck."

They also show up in tales and folklore, sometimes as gullible or foolish birds. The Blackfeet people, for example, tell of an old traveler who encounters a flock of ducks. The ducks implore the old man to stop and sing songs to them. He agrees on the premise that they dance to his songs with their eyes closed. As the man sings to the dancing birds, he snatches them up one by one and puts them into his cooking pot, with them being none the wiser. These waterfowl are depicted as dim-witted fools, but in many Native American cultures, ducks are sacred and respected symbols and their feathers are used in important ceremonial dances.

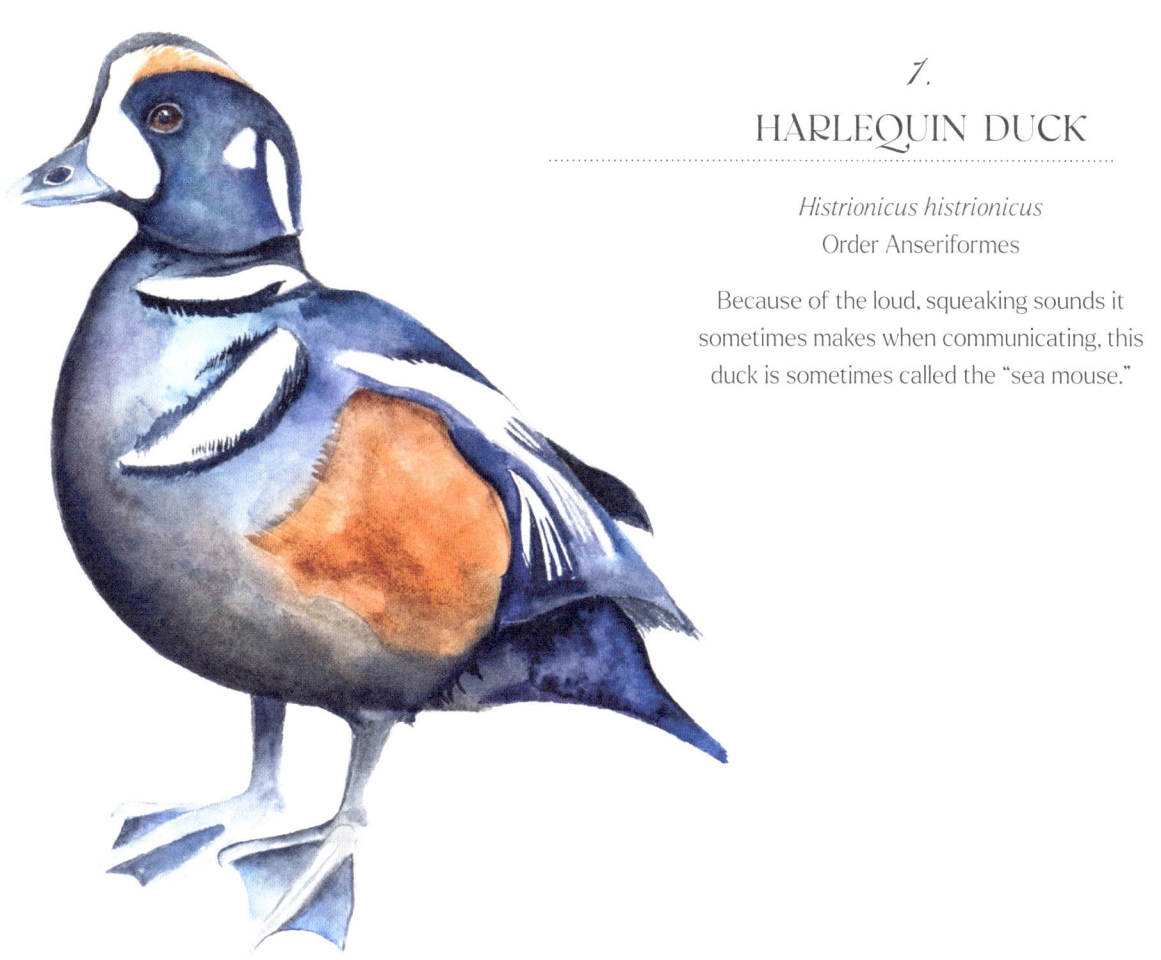

1.
HARLEQUIN DUCK

Histrionicus histrionicus
Order Anseriformes

Because of the loud, squeaking sounds it sometimes makes when communicating, this duck is sometimes called the "sea mouse."

2.
WOOD DUCK

Aix sponsa
Order Anseriformes

Wood ducks are one of the few waterfowl species that perch and nest in trees, where they use the sharp nails on their webbed feet to grip branches and maneuver into tree cavities. They often reuse nests created by woodpeckers.

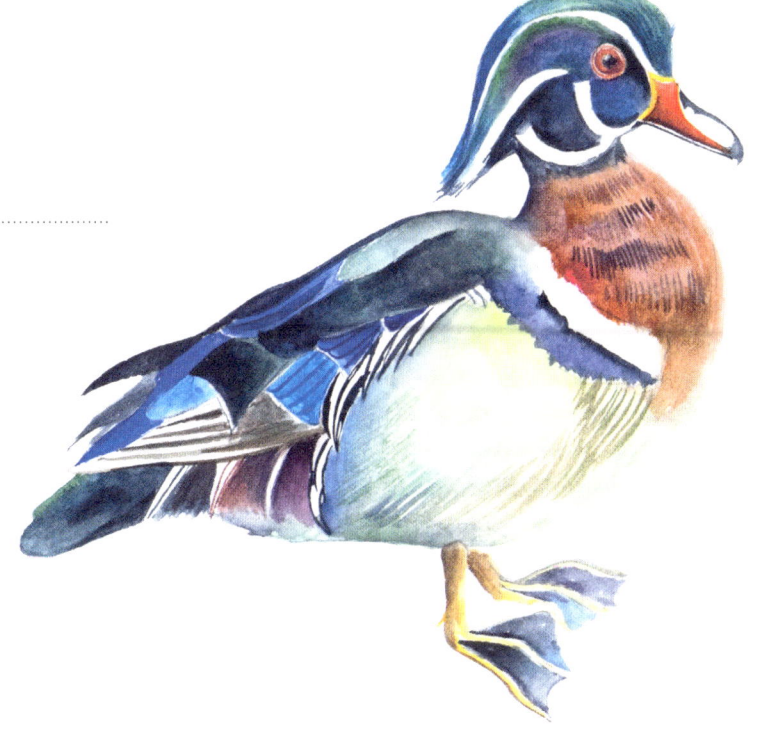

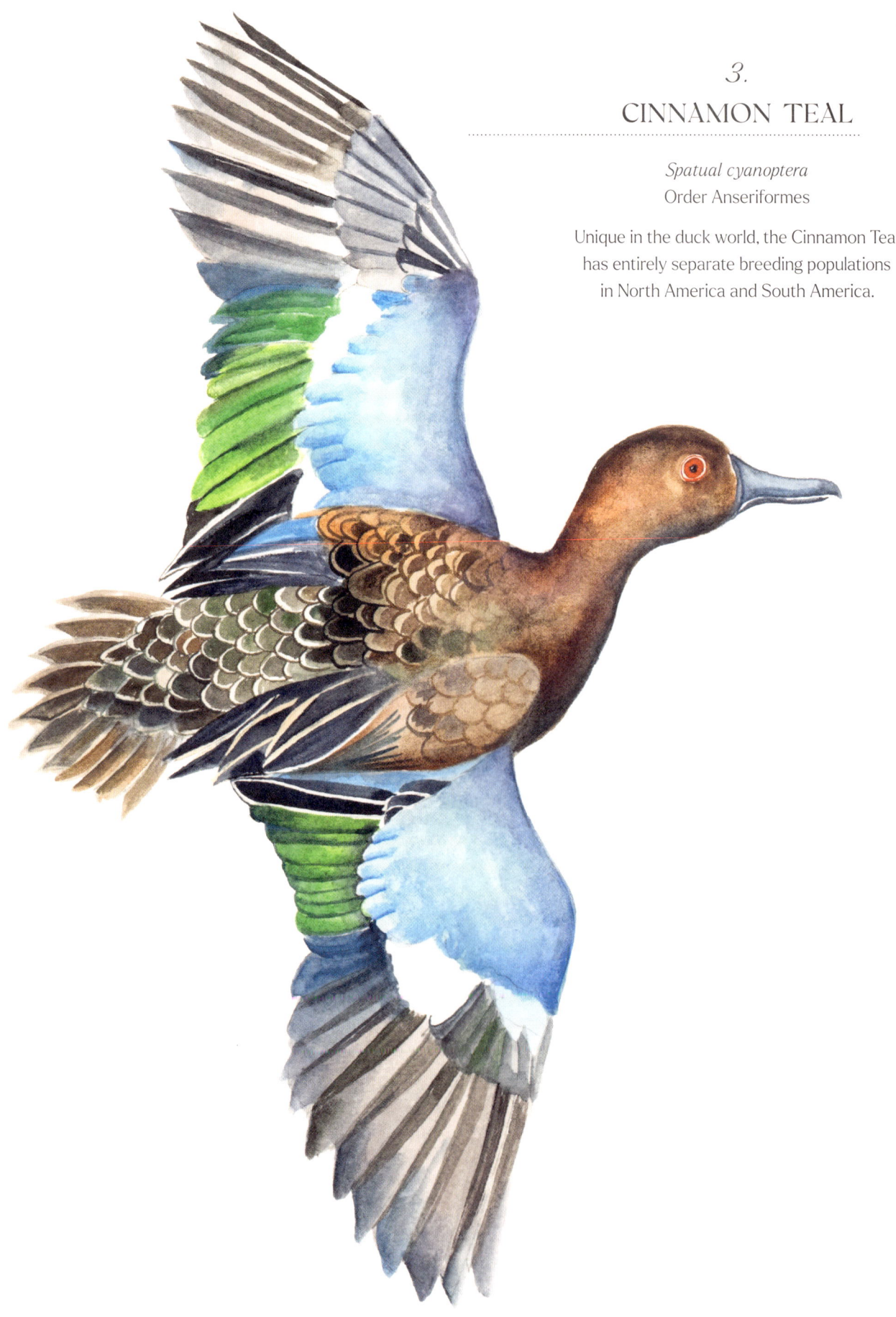

3.
CINNAMON TEAL

Spatual cyanoptera
Order Anseriformes

Unique in the duck world, the Cinnamon Teal
has entirely separate breeding populations
in North America and South America.

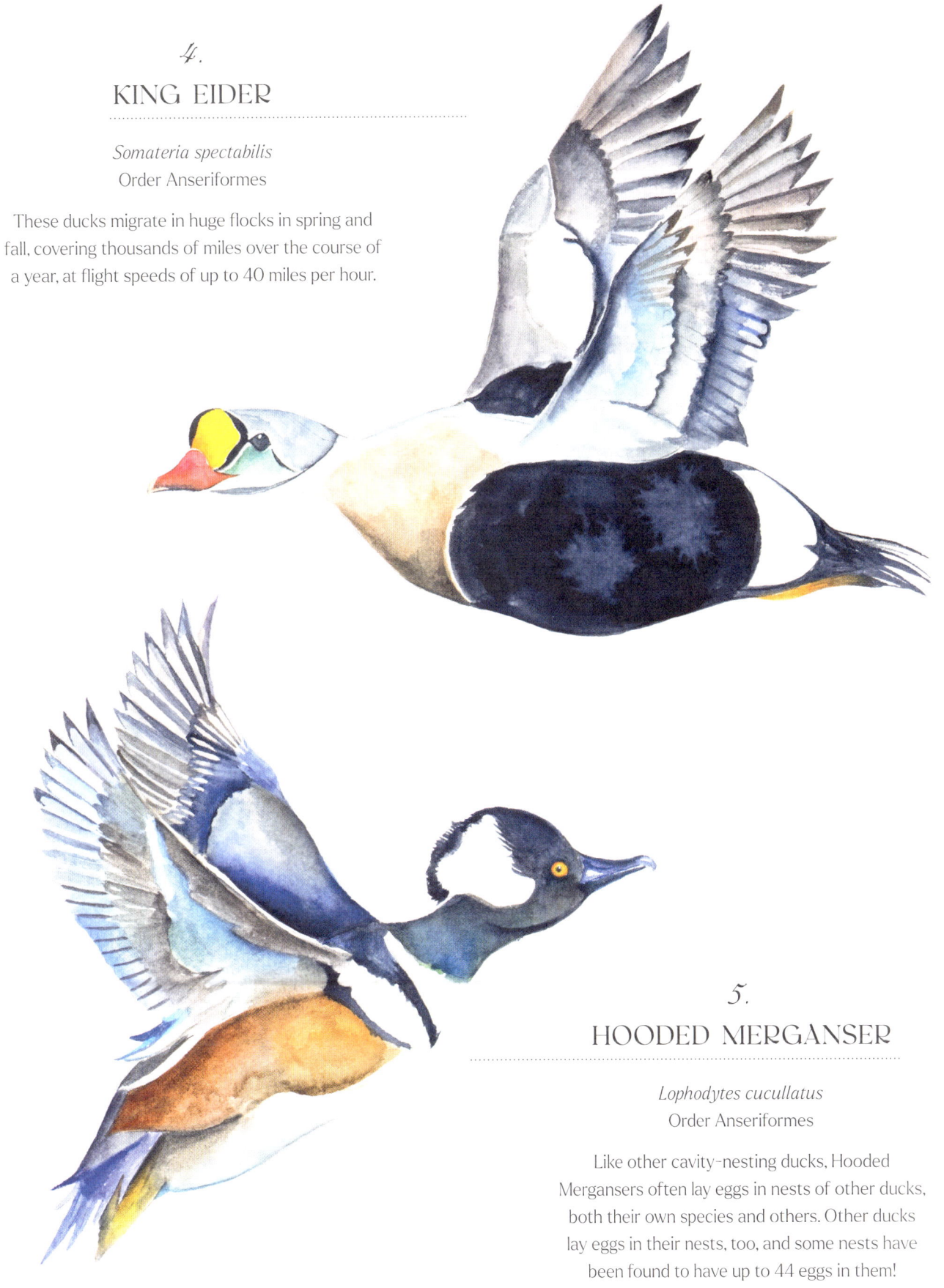

4.
KING EIDER

Somateria spectabilis
Order Anseriformes

These ducks migrate in huge flocks in spring and
fall, covering thousands of miles over the course of
a year, at flight speeds of up to 40 miles per hour.

5.
HOODED MERGANSER

Lophodytes cucullatus
Order Anseriformes

Like other cavity-nesting ducks, Hooded
Mergansers often lay eggs in nests of other ducks,
both their own species and others. Other ducks
lay eggs in their nests, too, and some nests have
been found to have up to 44 eggs in them!

OWLS

In folklore, owls are a pervasive symbol of magic, wisdom, mystery, and the occult. Most owl species are nocturnal, preferring to hunt under cover of darkness and sleep during the day. With their large, penetrating eyes, capable of seeing in the darkest recesses of the night, silent wings that carry them through the night unseen, and razor-sharp hearing that detects the smallest rustle of an animal in the dark, owls continue to mesmerize us. A common companion to sorcerers, shamans, and magic-makers, owls remind us of the unseen world that lies hidden beyond the crepuscular cusp of twilight.

Because of their mysterious nature, owls are viewed in several cultures as sinister omens. In some Native American traditions, they are considered omens of evil and harbingers of death and ill tidings. In the Pueblo tradition, owls are associated with the Skeleton Man, the god of death.

Conversely, the owl's association with death and the afterlife means to some that the birds hold the power to ward off other wicked powers. For example, in parts of China, owls are linked to thunder and lightning gods and are attached to rooftops to protect against lightning strikes. The Cherokee people believed that soaking owl feathers in water and placing the water over their eyes could enable them to stay awake all night long. Similarly in India, some believed that if you ate an owl's eyes, it would enable you to see in the night.

Owls are also often associated with wisdom and knowledge. The Greek goddess of wisdom, Athena, is often depicted with an owl companion perched on her shoulder. The owl is said to have passed on insights to her. The Little Owl (1) is often called "the Owl of Athena," as it appears in art and statues with the Greek goddess.

1.
LITTLE OWL

2.
BARN OWL

3.
EURASIAN EAGLE-OWL

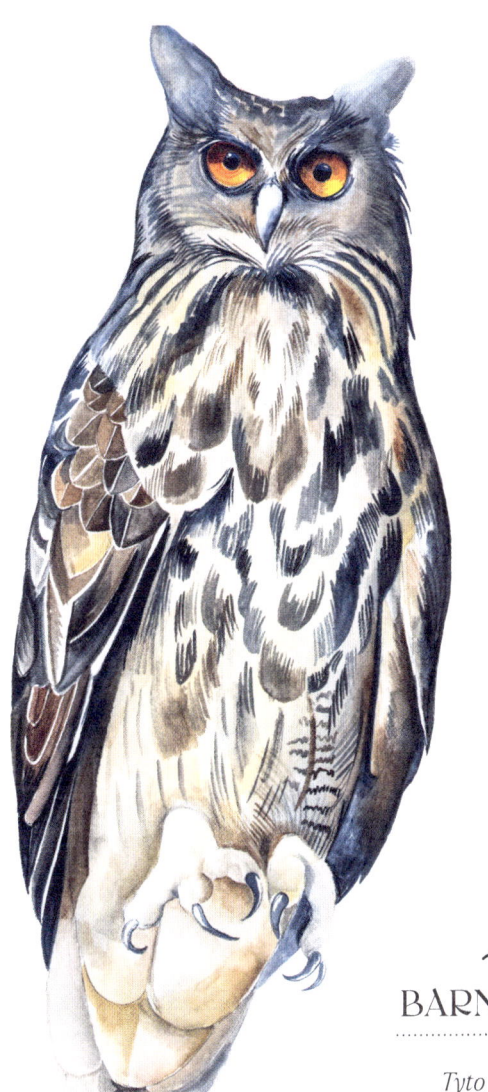

3.
EURASIAN EAGLE-OWL

Bubo bubo
Order Strigiformes

This owl was named for its formidable size. It
is the heaviest owl in the world, weighing 3 to
9 pounds, with a wingspan of 63 to 74 inches.

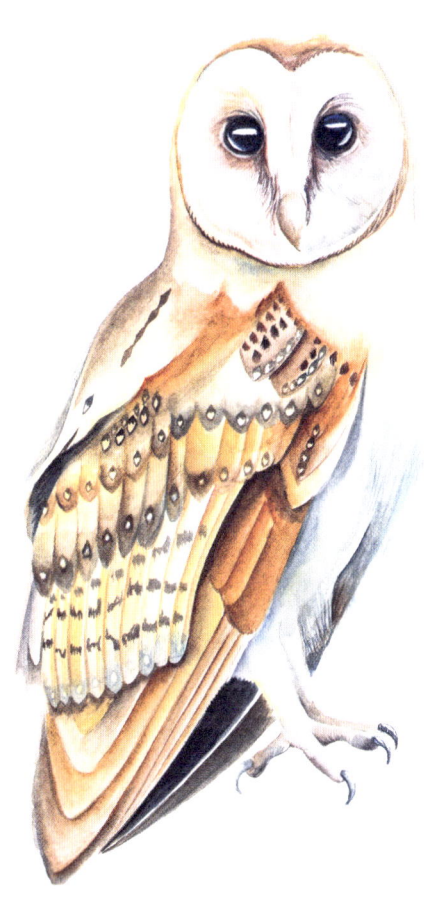

2.
BARN OWL

Tyto species
Order Strigiformes

Barn Owls are found on every continent of the world
except Antarctica. In 2024, ornithologists determined
that Barn Owls were of three similar species: the
American Barn Owl (*T. furcata*), the Western Barn
Owl (*T. alba*), and the Eastern Barn Owl (*T. javanica*).

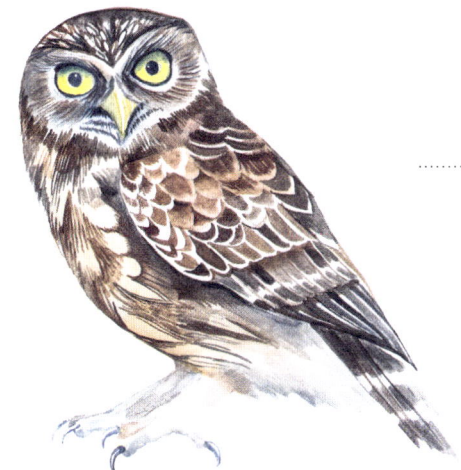

1.
LITTLE OWL

Athene noctua
Order Strigiformes

The "owl of Athena" has an extremely large range
that extends across Europe, Asia, and North Africa.
It lives in a variety of habitats including farmland,
woodland fringes, steppes, and semi-desert.

PHEASANTS

The word *pheasant* derives from the ancient Greek word *phāsiānós*, which means "the bird of Phasis." Phasis was the ancient Greek name of the Rioni River, east of the Black Sea. In Greek mythology, Jason and the Argonauts search for the Golden Fleece, which, if found, would restore him as heir to his father's throne. Some say that pheasants were one of the riches collected by Jason at Phasis.

The pheasant family (Phasianidae) includes quite a few stunning birds. One of the most well-known and beautiful family members is the peacock, a truly spectacular bird that deserves its cultural influence and reputation. Peacocks are native to the forests of Central and South Asia, but they have been introduced to many areas, where they are often considered nuisances. They are hunted worldwide for sport, and many cultures keep domesticated pheasants as display birds in gardens and aviaries. The males are at their most spectacular during mating season, when they strut back and forth and display their iridescent tail feathers.

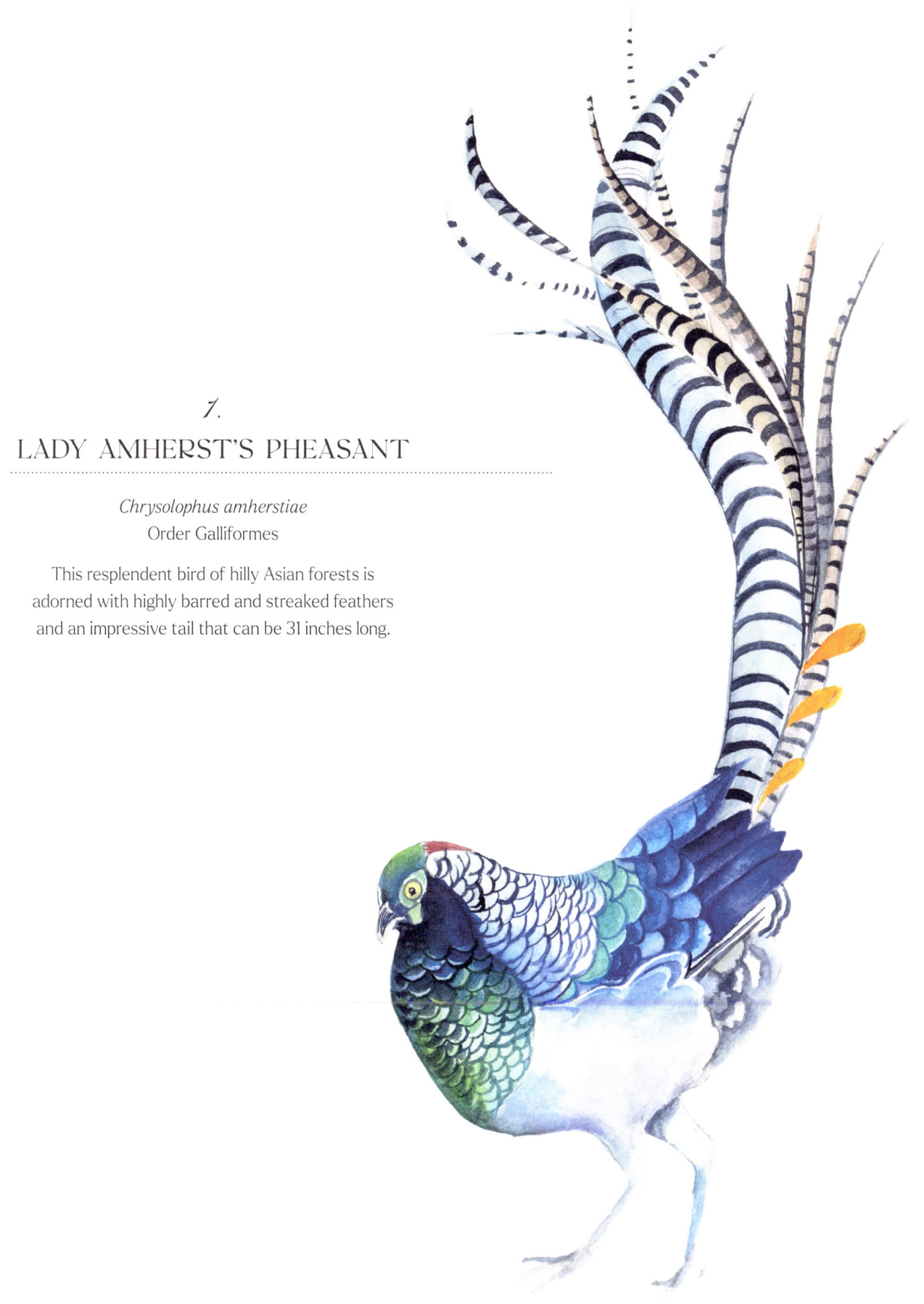

7.
LADY AMHERST'S PHEASANT

Chrysolophus amherstiae
Order Galliformes

This resplendent bird of hilly Asian forests is
adorned with highly barred and streaked feathers
and an impressive tail that can be 31 inches long.

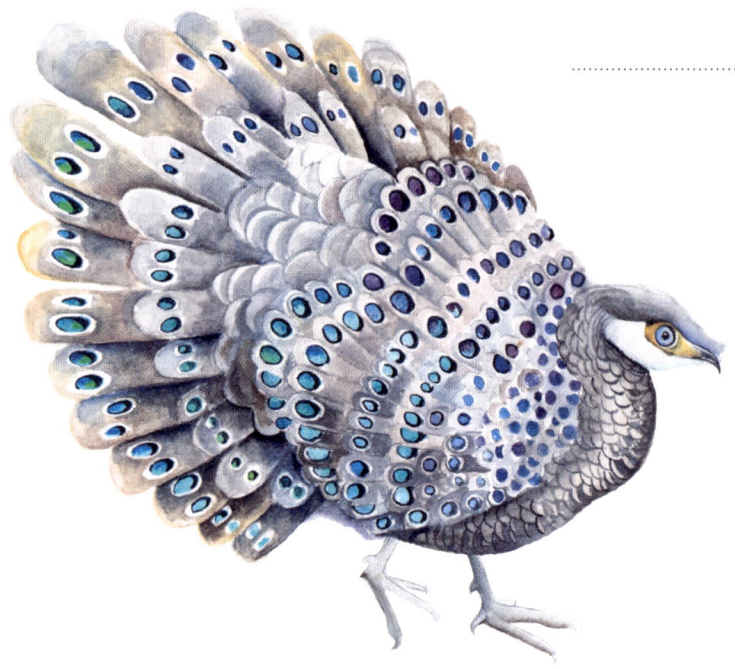

GRAY PEACOCK-PHEASANT

Polyplectron bicalcaratum
Order Galliformes

The male of this iridescently spotted pheasant opens its tail and spreads its wings like a peacock to attract a mate. The genus name is Ancient Greek for "many spurs" and refers to the several well-developed spurs on its legs that it uses for self-defense.

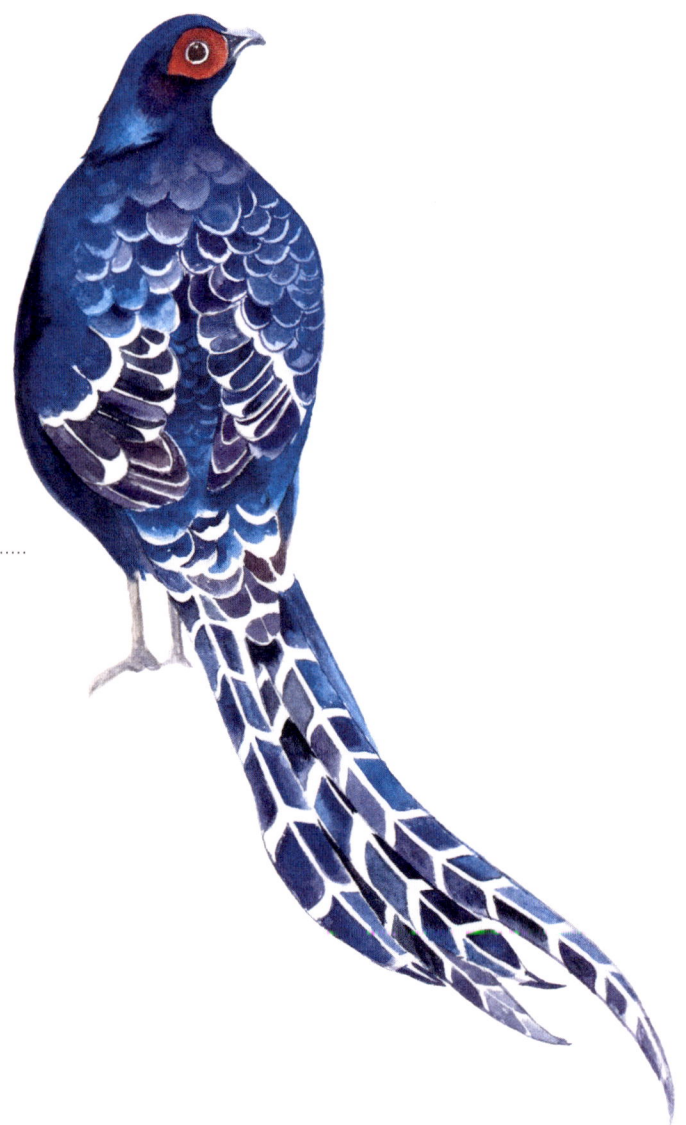

3.

MIKADO PHEASANT

Syrmaticus mikado
Order Galliformes

This gorgeous but threatened blue pheasant quietly forages on the ground for fruit, seeds, invertebrates, and vegetation in mountainous areas such as Yushan National Park, to which it is endemic.

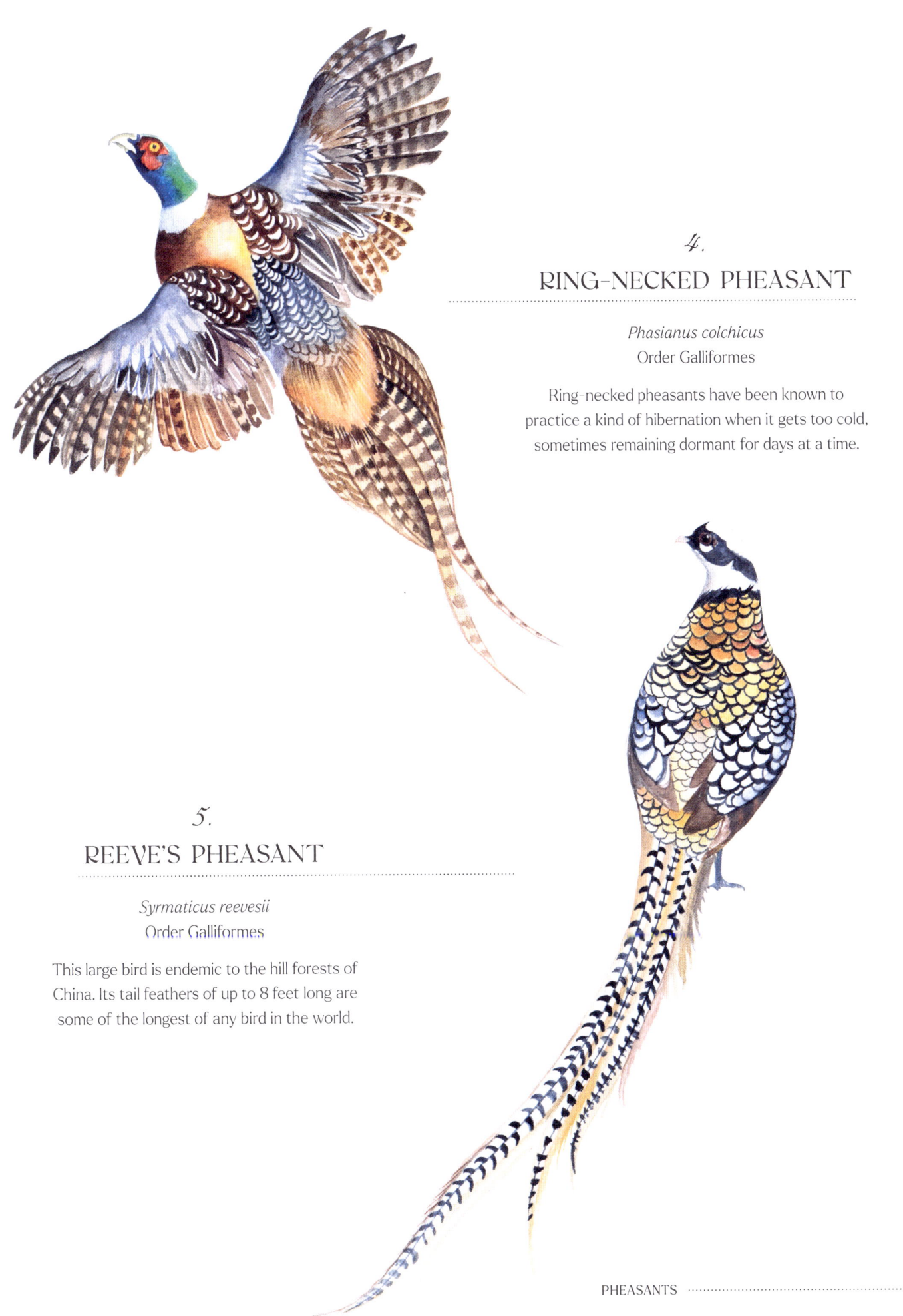

4.
RING-NECKED PHEASANT

Phasianus colchicus
Order Galliformes

Ring-necked pheasants have been known to
practice a kind of hibernation when it gets too cold,
sometimes remaining dormant for days at a time.

5.
REEVE'S PHEASANT

Syrmaticus reevesii
Order Galliformes

This large bird is endemic to the hill forests of
China. Its tail feathers of up to 8 feet long are
some of the longest of any bird in the world.

CHICKENS

Chickens are a favorite source of protein throughout the world, where people eat their eggs and meat in delicious recipes. Today, most of us think of chickens as domesticated birds. After all, humans have been raising and relying on domestic chickens for more than 4000 years. But what most of us don't realize is that all domestic chickens descended from the Red Junglefowl of Southeast Asia and Southwest China.

Hens are associated with fertility and domesticity in many cultures, and they serve as a perennial symbol for maternal care and attentiveness—the brooding hen, the mother hen, the nervous Little Red Hen in the children's story. Meanwhile, the rooster, often associated with the sun and dawn, serves as an enduring symbol of renewal, hope, and enlightenment. The crowing rooster greets the morning, and his red crest mimics the glowing red sky that announces the dawn. In Japanese Shinto mythology, the rooster is associated with Amaterasu, the goddess of the sun. In Chinese culture, the rooster symbolizes the sun, fidelity and punctuality, good fortune, protection, honesty, and spiritual guardianship. In ancient times, darkness was a fecund time for the activity of evil spirits and other nefarious creatures. The crowing of the rooster dispelled the potency of night and scattered the denizens and demons of the darkness to welcome the coming light of dawn.

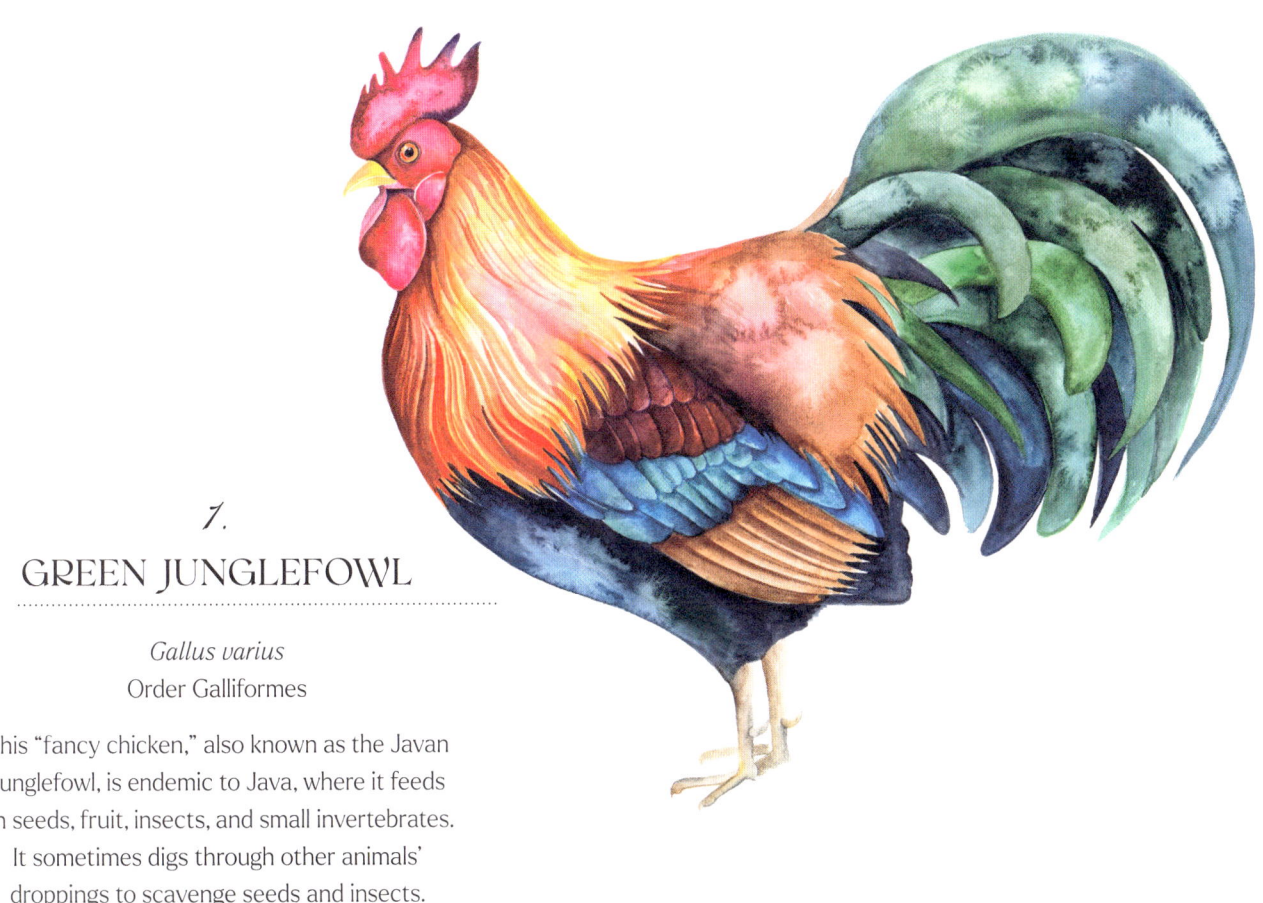

1.
GREEN JUNGLEFOWL

Gallus varius
Order Galliformes

This "fancy chicken," also known as the Javan Junglefowl, is endemic to Java, where it feeds on seeds, fruit, insects, and small invertebrates. It sometimes digs through other animals' droppings to scavenge seeds and insects.

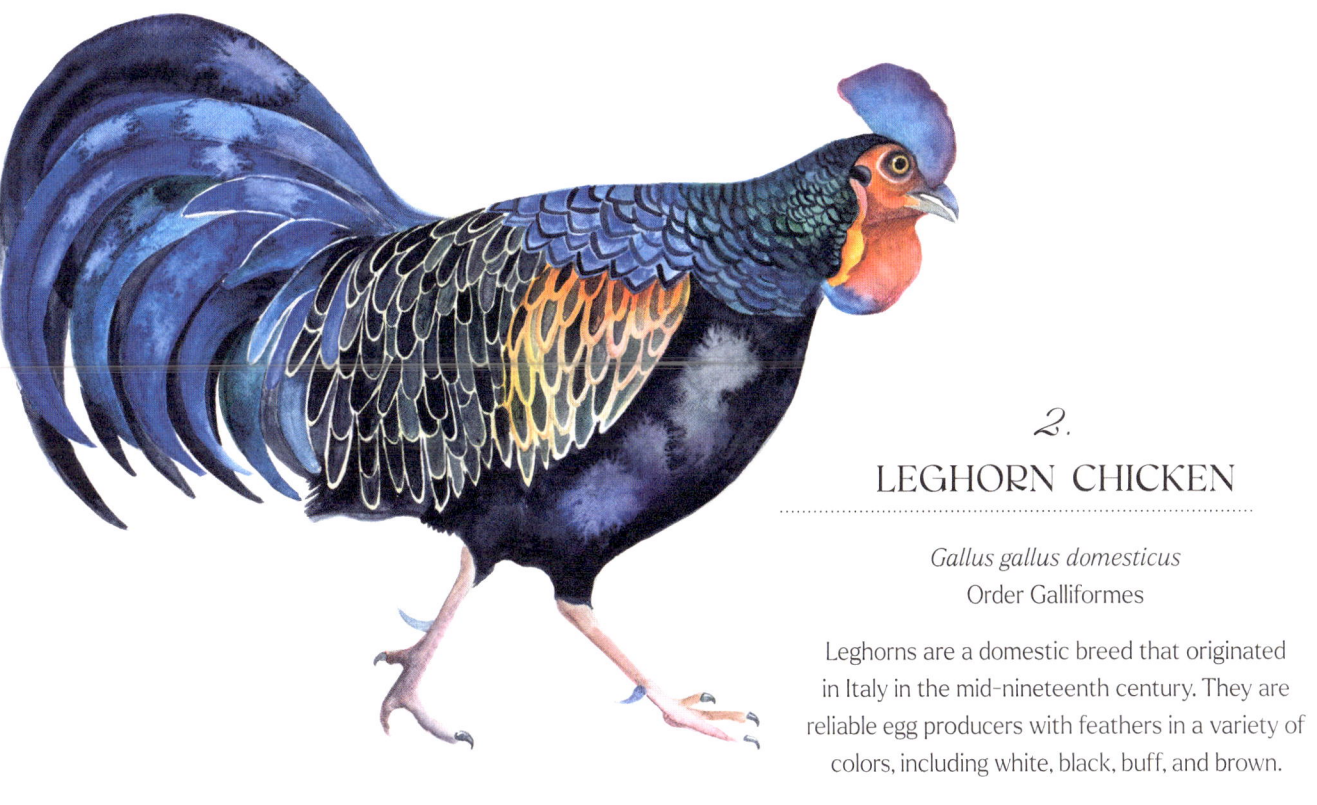

2.
LEGHORN CHICKEN

Gallus gallus domesticus
Order Galliformes

Leghorns are a domestic breed that originated in Italy in the mid-nineteenth century. They are reliable egg producers with feathers in a variety of colors, including white, black, buff, and brown.

RAPTORS

The word *raptor* comes from the Latin word *rapere*, which means "to seize" or "to take by force." Equipped with shrewd eyesight to spot unassuming prey from high above, razor-sharp talons for grabbing and killing prey, and piercing beaks to rip into flesh, this group of predatory birds represents the fiercest hunters of the bird world. Raptors are powerful hunters that live in many different habitats—from desert falcons that hunt above the shimmering heat of undulating sand dunes, to birds that seek prey near rivers, lakes, and other water bodies.

Raptors are often featured in folklore as hunters, messengers, tricksters, and sometimes bird royalty. In Native American tradition, hawks and eagles are seen as powerful symbols of courage and strength. The eagle is revered, and its feathers are often used as sacred objects. Birds of prey are also depicted in heraldic design as symbols of power. In Roman mythology, Jupiter, the god of the sky, used an eagle as his personal messenger that carried thunderbolts clasped in its talons.

American Kestrels (1), the smallest falcons in North America, inhabit many areas of open habitat. They sometimes perch on light posts during nighttime ballgames, snatching passing moths and other insects on the wing.

Although most raptors hunt and eat living vertebrates, the Crested Caracara (4), the national bird of Mexico, resembles a hawk and sometimes behaves like a vulture. It will eat carrion in addition to live prey, including turtles, snakes, fish, mammals, and other creatures. Though it looks like a hawk, this raptor is actually a falcon.

1.
AMERICAN KESTREL

2.
SECRETARY BIRD

3.
PHILIPPINE EAGLE

4.
CRESTED CARACARA

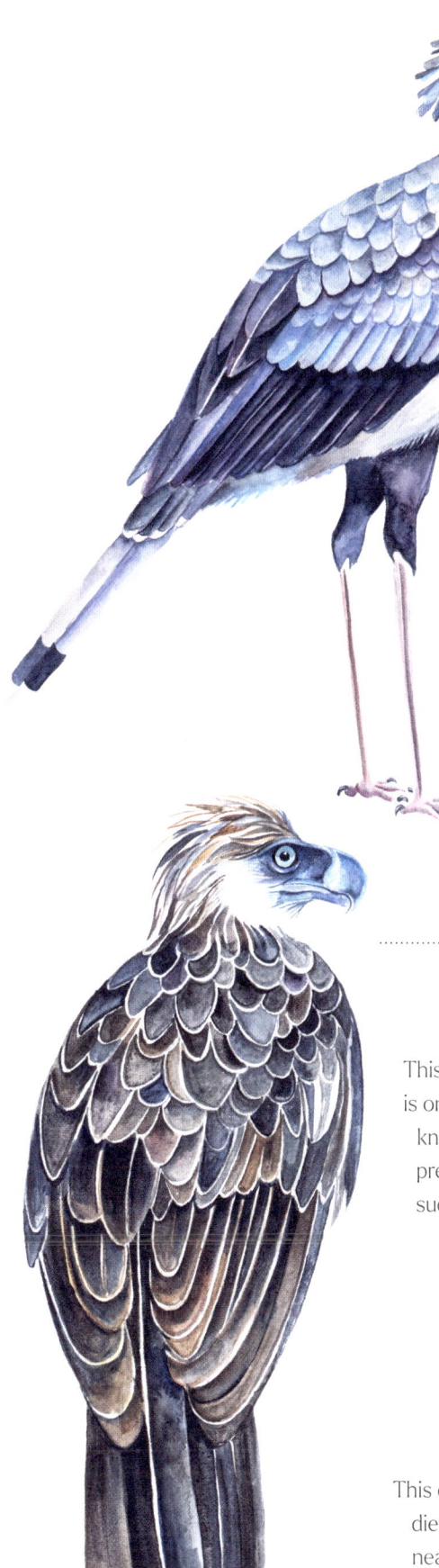

1.
AMERICAN KESTREL

Falco sparverius
Order Falconiformes

Unlike humans, kestrels can see ultraviolet light, which makes it possible for them to track prey by looking for trails of their urine on the ground.

2.
SECRETARY BIRD

Sagittarius serpentarius
Order Accipitriformes

This long-legged raptor strides through the African savanna, hunting for reptiles, small mammals, and insects. It kills its prey by bludgeoning them with its legs. Its Latin species name means "serpent handler" and may refer to its skill at hunting snakes.

3.
PHILIPPINE EAGLE

Pithecophaga jefferyi
Order Accipitriformes

This critically endangered species of the Philippines is one of the largest eagles in the world. It was once known as "the monkey-eating eagle" because its prey includes primates as well as other mammals such as lemurs, small deer, pigs, and young goats.

4.
CRESTED CARACARA

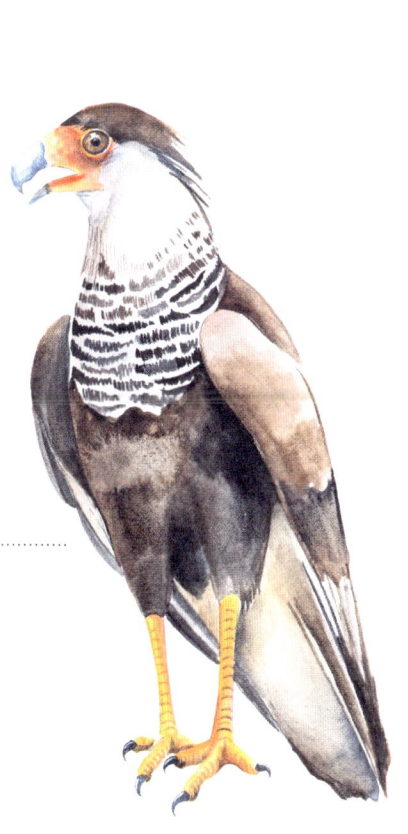

Caracara plancus
Order Falconiformes

This opportunistic feeder has one of the most diverse diets of any species of diurnal raptor. It consumes nearly anything, from carrion to tiny ants, and will plunder the nests of other birds, even other caracaras.

FLIGHTLESS BIRDS

Birds are perennial symbols of flight and freedom, and they soar high above the earth on wings, but several birds do not possess the ability to fly. One of the most recognizable flightless birds is the Common Ostrich (1), which is also the heaviest of all living birds. Ostriches can weigh up to nearly 300 pounds and reach 9 feet tall, and they can run up to 45 miles per hour. Their large and beautiful feathers have been used for ornamentation on hats and clothing, and their eggs and meat are consumed in many parts of the world as a delicacy. Contrary to popular belief, ostriches do not bury their heads in the sand when they feel threatened. They do, however, lay their eggs in shallow holes in the sand and use their beaks to turn the eggs during the incubation process.

Another fascinating and large flightless bird is the Southern Cassowary (2), native to the rainforests of Australia, New Guinea, and Indonesia. With its technicolored reptilian countenance, formidable height, powerful legs and feet, and daggerlike toenails, this distinctive bird is unmistakable and has been called "the world's most dangerous bird." If a cassowary feels threatened, it may attack, and even kill, though attacks on humans are rare. Unprovoked, however, it is a shy and peaceful bird.

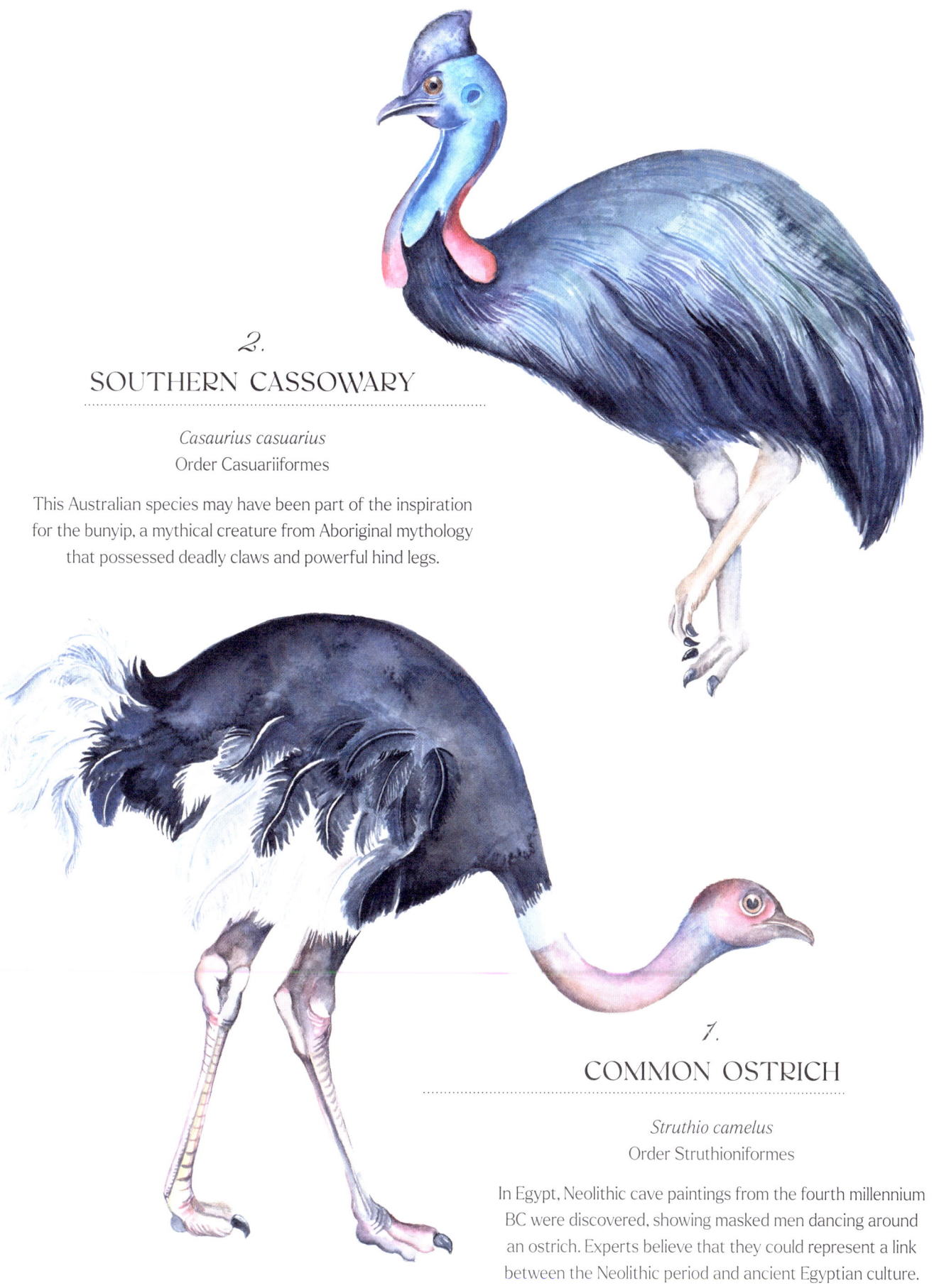

2.
SOUTHERN CASSOWARY

Casaurius casuarius
Order Casuariiformes

This Australian species may have been part of the inspiration for the bunyip, a mythical creature from Aboriginal mythology that possessed deadly claws and powerful hind legs.

1.
COMMON OSTRICH

Struthio camelus
Order Struthioniformes

In Egypt, Neolithic cave paintings from the fourth millennium BC were discovered, showing masked men dancing around an ostrich. Experts believe that they could represent a link between the Neolithic period and ancient Egyptian culture.

CORVIDS

This group of highly intelligent birds includes crows, ravens, and magpies. Corvids are often associated with knowledge, bad luck, ill omens, and mischief making. Opportunistic hunters, ravens and crows are omnivores that will eat almost anything that is available, including carrion.

Morrigan, a goddess of war, fate, and death in Celtic mythology, was said to transform into a raven or crow to fly over the battlefield looking for corpses to feed on, a mythology no doubt inspired by the real habits of wild ravens. Danish folklore tells of a valravn, a supernatural raven that consumes the bodies of the dead and gains human knowledge and intelligence from the flesh of the departed souls. More flatteringly, in Norse mythology, the powerful god Odin entrusted two loyal ravens, Muninn and Huginn, to deliver him the hidden truths of the world. The raven is the first bird mentioned by name in the Bible: in Genesis, a raven is named as the first animal released from Noah's ark.

Magpies, including the Eurasian Magpie (2), are also featured in mythology. In British folklore, the magpie was seen as a harbinger of bad luck. Saluting a magpie is said to keep that bad luck at bay. Some people even cross themselves when they see a magpie. But some folklore views magpies in a more positive light. In China and Korea, for example, these birds are believed to bring good luck and are symbols of happiness. Magpies have been unfairly maligned as thieves that gather shiny objects and store them in their nests, but scientists have found no evidence to support this.

1.
COMMON RAVEN

2.
EURASIAN MAGPIE

3.
HOODED CROW

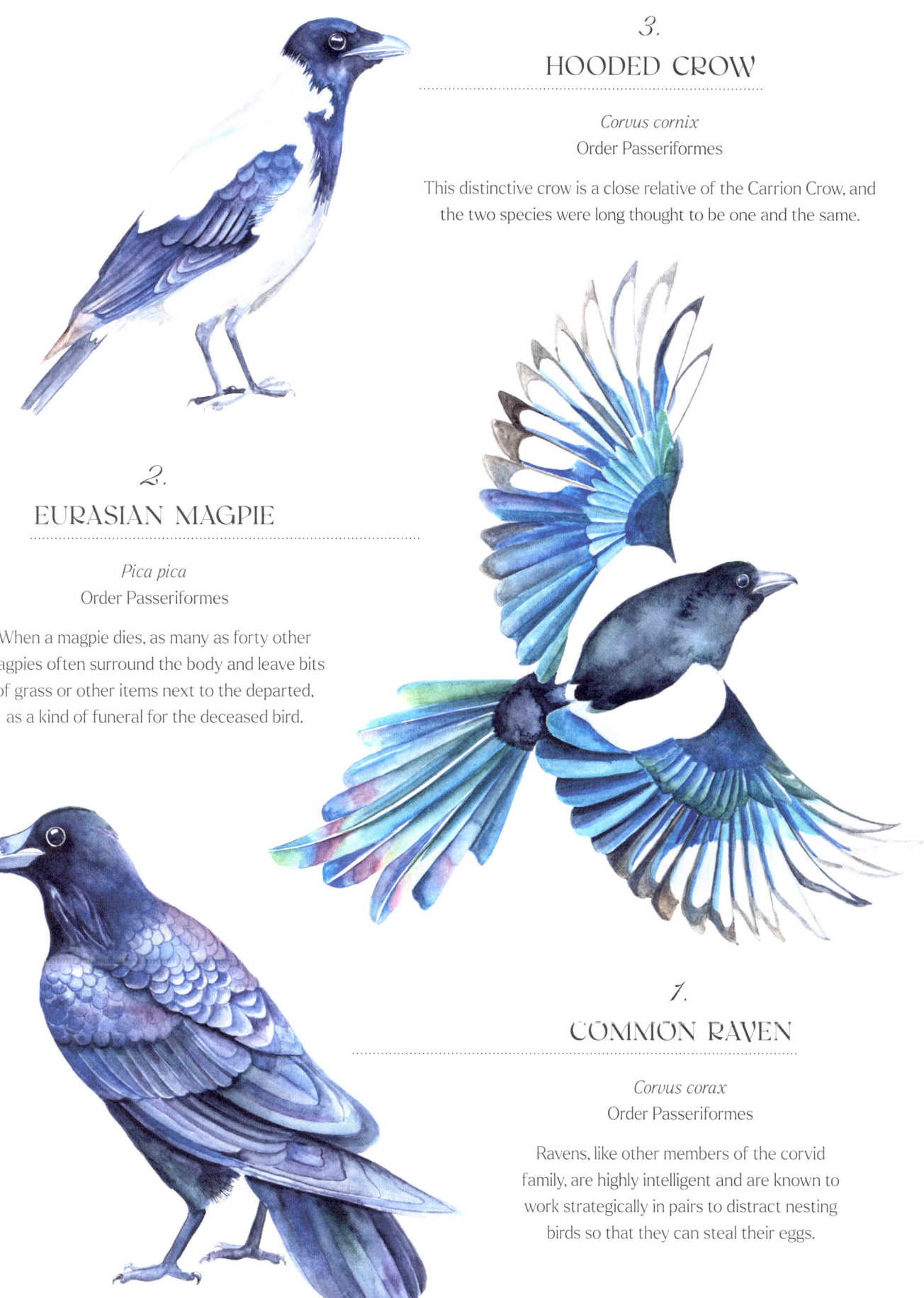

3.
HOODED CROW

Corvus cornix
Order Passeriformes

This distinctive crow is a close relative of the Carrion Crow, and the two species were long thought to be one and the same.

2.
EURASIAN MAGPIE

Pica pica
Order Passeriformes

When a magpie dies, as many as forty other magpies often surround the body and leave bits of grass or other items next to the departed, as a kind of funeral for the deceased bird.

1.
COMMON RAVEN

Corvus corax
Order Passeriformes

Ravens, like other members of the corvid family, are highly intelligent and are known to work strategically in pairs to distract nesting birds so that they can steal their eggs.

PIGEONS

Many of us are quite familiar with the Rock Pigeon (4). This ubiquitous avian nuisance flutters through cities and forages on sidewalks around the world, scavenging trash and edible detritus. These birds, however, deserve quite a bit more respect than they are afforded in contemporary life. Their successful integration into urban spaces is a sign of their great intelligence and adaptability. They are incredibly sophisticated navigators and can find their way home from at least 2500 miles away using a precise internal GPS-like system. Because of this skill, pigeons have been used as messengers by various cultures for thousands of years. In fact, the Rock Pigeon we know today is a descendant of pigeons domesticated in the Middle East many millennia ago. Hieroglyphics from ancient Egypt even suggest that the Egyptians domesticated pigeons, using them as messengers and for food.

Colorful variations in the pigeon family include the Luzon Bleeding-heart (3), named for the crimson patch across its chest. Native to the Philippines, these beautiful birds live in the rainforest, where they forage on the ground for seeds and insects. The Nicobar Pigeon (2) is a vibrant, psychedelically colored pigeon from Nicobar Islands, a small chain of islands in Southeast Asia. These pigeons are believed to be the closest living relatives of the long-lost dodo of Madagascar.

1.
SPINIFEX PIGEON

Geophaps plumifera
Order Columbiformes

Sometimes known as the "plumed pigeon" for its spiky crest, this Australian endemic forages on the ground in arid grasslands dominated by spinifex, a tussock-forming grass that grows on sandy hills.

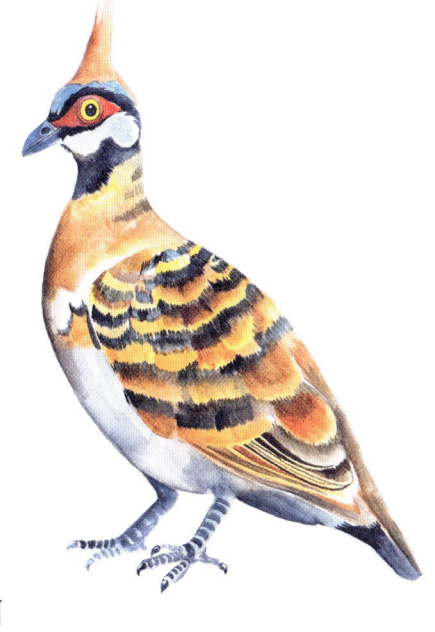

2.
NICOBAR PIGEON

Caloenas nicobarica
Order Columbiformes

This large and beautiful pigeon from Southeast Asia is the only living member of its genus. Its dark, iridescent plumage hides the bright white tail feathers that are difficult to see unless the bird is in flight.

3.
LUZON BLEEDING-HEART

Gallicolumba luzonica
Order Columbiformes

Because of the bright red feather patch on its chest, the Tagalog name for this forest-dwelling bird of Southeast Asia is *paloma de punalada*, meaning "stabbed pigeon."

4.
ROCK PIGEON

Columba livia
Order Columbiformes

Because pigeons can travel great distances, some cultures consider them symbolic messengers between divine realms. Others consider them pests, or "rats with wings," because of their scavenging tendencies and ubiquity throughout urban centers.

PENGUINS

These flightless seabirds live primarily in the Southern Hemisphere. They are often associated with Antarctica, but several penguin species live in milder climates, including Australia and New Zealand. Only one species of penguin, the Galápagos Penguin, lives north of the Equator. The Great Auk, now extinct, was once considered the first "penguin," though these two birds are now known to be distinct species. Penguins have endeared themselves to humans for many years with their unusual waddling walk and distinctive tuxedo-like markings. Because most penguins have evolved in places with few natural predators, they are often unafraid of humans.

At nearly 40 inches tall, the King Penguin (2) is the second largest penguin, after the Emperor Penguin. They are gregarious birds, particularly during mating season, and can nest in colonies that number thousands of pairs. Each female lays a single egg and does not build a nest; instead, parents take turns carrying the egg on their feet for the 55-day incubation period. When an egg hatches, the fuzzy brown penguin chick is also carried on its parents' feet for a month or so until it develops enough plumage to regulate its body temperature. Young King Penguins look so different from adults that when early explorers first encountered them, they believed them to be two entirely separate species of birds.

The much smaller Little Penguin (3) is native to New Zealand and is one of the smallest penguins in the world, at about 10 inches tall. Because of their petite size and inability to fly, they are vulnerable to attack by introduced mammal predators in New Zealand. Even household pets like dogs and cats have been known to prey on these seabirds. Because of this, more and more of them nest on islands off the shores of New Zealand, where they are less imperiled by other animals.

1.
MAGELLANIC PENGUIN

2.
KING PENGUIN

3.
LITTLE PENGUIN

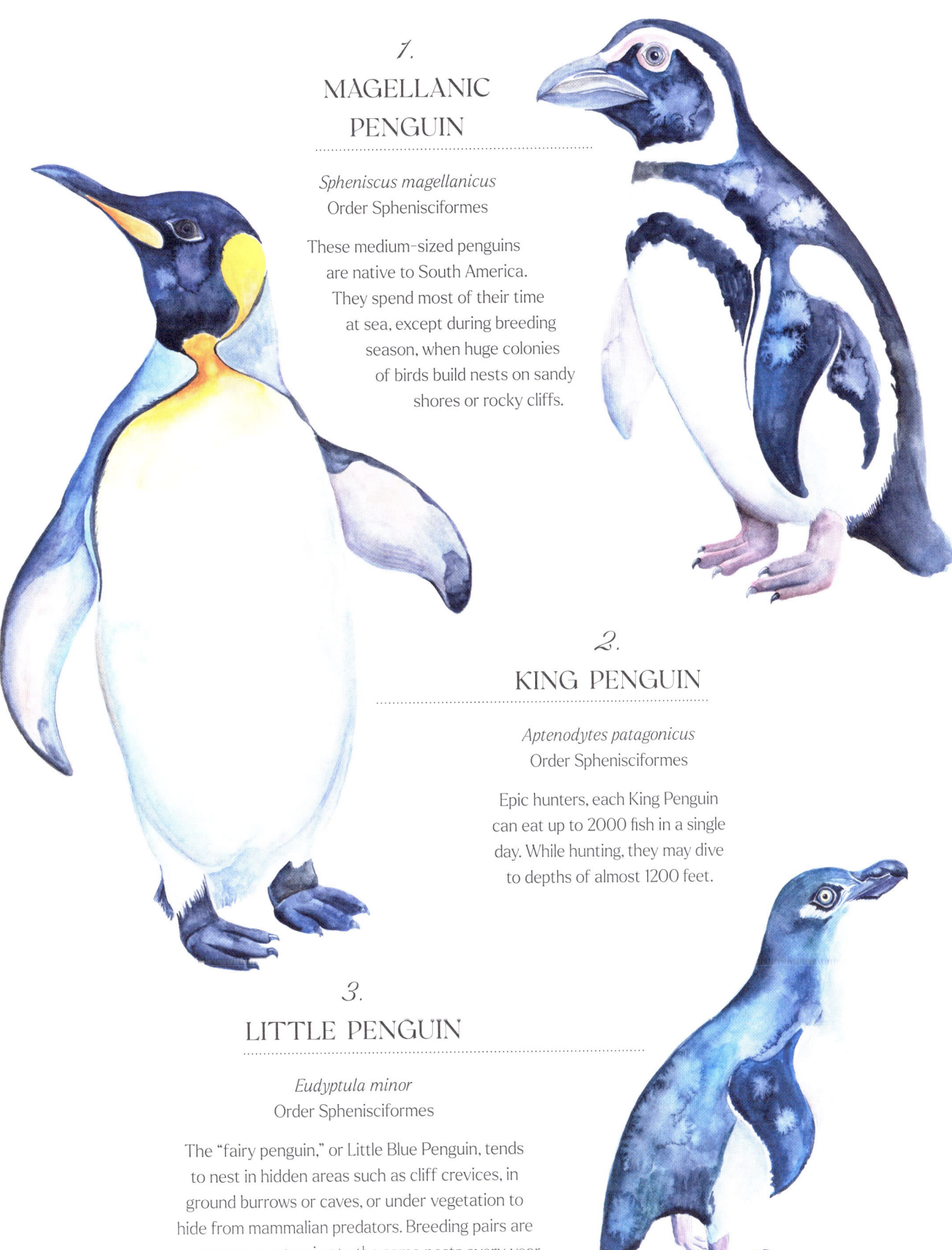

1.
MAGELLANIC
PENGUIN

Spheniscus magellanicus
Order Sphenisciformes

These medium-sized penguins
are native to South America.
They spend most of their time
at sea, except during breeding
season, when huge colonies
of birds build nests on sandy
shores or rocky cliffs.

2.
KING PENGUIN

Aptenodytes patagonicus
Order Sphenisciformes

Epic hunters, each King Penguin
can eat up to 2000 fish in a single
day. While hunting, they may dive
to depths of almost 1200 feet.

3.
LITTLE PENGUIN

Eudyptula minor
Order Sphenisciformes

The "fairy penguin," or Little Blue Penguin, tends
to nest in hidden areas such as cliff crevices, in
ground burrows or caves, or under vegetation to
hide from mammalian predators. Breeding pairs are
monogamous, returning to the same nests every year.

IN MEMORIAM:
EXTINCT BIRDS

Human behaviors and climate change have been responsible for the degradation and eradication of various bird species around the world. The Wildlife Society estimates that some 1400 bird species have become extinct throughout modern human history. Here you'll learn about three of the many birds that have been lost to our world.

The Passenger Pigeon (1) was a wild pigeon endemic to North America that was once the most abundant bird on the continent, numbering in the billions. These birds were a common sight in the forests of eastern North America and were revered and respected by various Indigenous tribes, some of whom believed the birds were the reincarnated souls of the departed. But American settlers considered them pests, as large migrating flocks damaged crops, and because the pigeons were a plentiful source of food, they were hunted in great numbers. By the mid-nineteenth century, tens of millions of birds had been shot for sport, and the last recorded wild bird was shot in 1900. When the last captive bird died in 1914 at the Cincinnati Zoo, so did the species. Its extinction spurred the Lacy Act of 1900, the first law to protect wildlife in the United States.

Today it might be difficult to imagine parrots or parakeets in the forests of eastern North America, but the Carolina Parakeet (2) was one such parrot that was tragically driven to extinction years ago. Like Passenger Pigeons, these birds were seen as crop predators, and it was common practice for farmers to shoot them to protect their harvest. They were also prized for their brightly colored tail feathers, which were a popular adornment on women's hats. These factors, combined with major habitat loss, drove them to total extinction in the early twentieth century.

The Ivory-billed Woodpecker (3), the largest species of woodpecker in North America, was considered extinct in the early twentieth century until a sighting in Louisiana in 1944 changed its status to critically endangered. Habitat destruction and deforestation are among the factors that led to the decline of these beautiful birds, and today they are once again considered extinct. The last credible sighting was noted in Arkansas in 2004.

1.
PASSENGER
PIGEON

2.
CAROLINA
PARAKEET

3.
IVORY-BILLED
WOODPECKER

3.

IVORY-BILLED WOODPECKER

Campephilus principalis
Order Piciformes

Native Americans used these birds' bills for decorations and for trade with other tribes. Archeologists have found Ivory-billed Woodpecker skulls far outside of the woodpecker's known range.

2.

CAROLINA PARAKEET

Conuropsis carolinensis
Order Psittaciformes

Last seen in the wild in the early twentieth century, these neotropical parakeets had the northernmost range of any known parrot. They once traveled in large, raucous flocks that gorged themselves on seeds and fruits.

1.

PASSENGER PIGEON

Ectopistes migratorius
Order Columbiformes

These birds once followed a nomadic lifestyle, flying in enormous flocks to seek out bumper crops of acorns and beechnuts produced by oak and beech trees.

Artist's Note

I grew up in rural New England, waking up to robins hopping merrily outside my window, spending my summer days searching for wild blackberries in the Berkshire hills, and waiting for the Great Blue Heron to erupt from the trees along the river where I grew up. I spent my childhood outside, and I developed an abiding fascination with birds, plants, and the natural world.

My artwork has long been inspired by themes of nature, climate change, and biodiversity, and research has always been a vital component of my practice. My work as an artist has grown out of my ongoing fascination with natural history, mythology, and the ways in which humans interact with the natural world. To make my paintings, I spend a great deal of time poring over nature books looking for the facts, images, and stories that inspire their creation. Through this process, I learn and connect to the elements I add to the work. For a long time, much of this research never saw the light of day. Scribbles of birds and scrawls of little

bird factoids that I found inspiring have lived only in my sketchbooks. With the encouragement of a few friends, I came to realize that perhaps this collection of research could become its own work; thus, the idea for *The Illuminated Book of Birds* was born.

For me, this project has been an opportunity to create a book that grows both out of and into my art practice. It is a chance to share the research I have long conducted privately and for me to deepen and continue that research as I create the book.

In our contemporary world, where images and things are designed and produced en masse to be consumed, I believe there is a politic in working by hand, with extreme attention to detail. I consider it a privilege and an act of quiet rebellion to make these images slowly, deliberately, and by hand. I hope you take time with them and with the natural world to reveal the things not present to you with just a cursory glance.

Acknowledgments

I would like first to thank my wonderful husband, Jake Brodsky. He has patiently listened to me go on and on about this project and many others for years, collaborated on various art projects with me, and made time and space for me to work in my studio uninterrupted. Without you, none of this would be possible. Thank you to my son, Felix, whose squeals of delight when he sees a new bird make this work feel even more worth doing.

I am hugely indebted to my dear friend and fellow writer Mike Macdonald, without whom you would probably not be reading this book today. Mike was one of the first people to believe this book was worthy of being put out in the world. He has helped me along every step of the way, from patiently reading my very first proposal drafts, to letting me write for days at his kitchen table as my deadline approached. He has pushed me to think bigger and aim higher every step of the way, and this is a better book for all his advice and guidance.

Thank you to my agent, Dawn Frederick, for taking a chance on me, for patiently answering all my questions, and for finding this book such a perfect home at Timber Press. Thank you everyone at Timber Press for believing in this book and working so hard to make it come to life. The vision, guidance, design, and editing of the Timber team have made this a much better book.

This project would not have been possible without the expertise and devotion of all those people who spend their lives studying, understanding, and protecting birds. I would like to thank all the much more knowledgeable scientists and researchers to whom I looked for inspiration and information when writing and researching this book.

Thank you to all my friends who have supported me and my art over the years, who have attended my openings, shared and collected my art, talked over book ideas with me at the dog park, and encouraged me to keep going.

Index

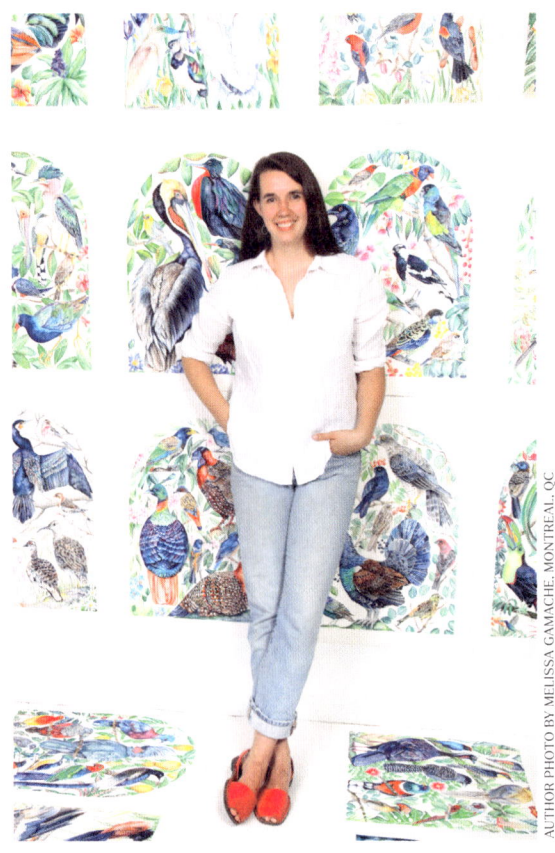

AUTHOR PHOTO BY MELISSA GAMACHE, MONTREAL, QC

ROBIN CROFUT-BRITTINGHAM is an artist and writer. Her practice focuses on the exploration of real and imagined worlds, inspired by nature. Originally from Western Massachusetts, she received her BA in Writing from Bard College and her MFA from the San Francisco Art Institute. Her work has been featured in many international publications and exhibited in solo and group exhibitions throughout the United States and Canada. Her paintings and murals are included in collections around the world. She lives and works in Montreal. To see more of Robin's work visit www.robincb.com or Instagram @robincbstudio.